NUREG-1609
Supplement 2

Standard Review Plan for Transportation Packages for Irradiated Tritium-Producing Burnable Absorber Rods (TPBARs)

I0502852

Manuscript Completed: June 2005
Date Published: February 2006

R.W. Parkhill, NRC Project Manager

Spent Fuel Project Office
Office of Nuclear Material Safety and Safeguards
U.S. Nuclear Regulatory Commission
Washington, DC 20555-0001

ABSTRACT

The NRC contracted with LLNL to compile this supplement to NUREG-1609 to incorporate additional information specific to tritium-producing burnable absorber rods (TPBARs). As a supplement to NUREG-1609, this report is intended to provide details on transportation package review guidance for the shipment of TPBARs. The principal purpose of this supplement is to ensure the quality and uniformity of staff reviews of packagings intended for transport of TPBARs. It is also the intent of this plan to make information about regulatory matters widely available, and improve communications between the NRC, interested members of the public, and the nuclear industry, thereby increasing the understanding of the NRC staff review process. In particular, this supplemental guidance, together with NUREG-1609, assists potential applicants by indicating one or more acceptable means of demonstrating compliance with the regulations.

CONTENTS

FIGURES

TABLES

ABBREVIATIONS AND ACRONYMS

Code of Federal Regulations	CFR
Hypothetical Accident Conditions	HAC
Interim Staff Guidance	ISG
Maximum Normal Operating Pressure	MNOP
Normal Conditions of Transport	NCT
Nuclear Regulatory Commission	NRC
Regulatory Guide	RG
Safety Analysis Report	SAR

INTRODUCTION

The *Standard Review Plan for Transportation Packages for Radioactive Material* (NUREG 1609)[1] provides guidance for the U.S. Nuclear Regulatory Commission's (NRC's) safety reviews of packages used in the transport of radioactive materials (RAM) under Title 10 of the U.S. Code of Federal Regulations (CFR), Part 71 (10 CFR Part 71). It is not intended as an interpretation of NRC regulations. NUREG-1609 supplements NRC Regulatory Guide (RG) 7.9, *Standard Format and Content of Part 71 Applications for Approval of Packaging for Radioactive Material*,[2] for review of package applications. NUREG-1609 involves guidance for reviewing radioactive material packagings intended for transport of a variety of radioactive materials, with the exception of spent nuclear fuel. Comparable guidance for the transport of spent fuel can be found in NUREG-1617, the *Standard Review Plan for Transportation Packages for Spent Nuclear Fuel.*[3]

The current report is not a stand-alone document, but is intended primarily as a supplement to NUREG-1609. It should also be noted, however, that, in some ways, this report can also be considered as a supplement to NUREG-1617. As a supplement to NUREG-1609, this report is intended to provide details on package review guidance for the shipment of irradiated Tritium-Producing Burnable Absorber Rods (TPBARs). During the irradiation process, TPBARs function in the reactor core like any other burnable poison rods, with the notable exception that TPBARs are designed to produce tritium. Thus, on the one hand, the primary purpose of this document is to provide guidance for the review of tritium shipping containers. On the other hand, however, because TPBARs function in the reactor core like any other burnable poison rods, the shipment of irradiated TPBARs can be expected to take on all of the shielding considerations of shipping containers for spent nuclear fuel, without having to deal with any of the criticality concerns.

As a supplement to NUREG-1609, this report is organized in the same manner as NUREG-1609, and has the identical numbering of subsections as found in that document. In addition, the appendices found in this supplement are labeled to allow this report to be completely merged with NUREG-1609, and subsequent supplements, without the need to change any of the labeling. For example, NUREG-1609 had two appendices labeled A and B, with Appendix A being composed of eight parts. An earlier supplement to NUREG-1609,[4] that provided guidance on considerations for unirradiated MOX fuels, contained four appendices, with two labeled A-9 and A-10, and the other two labeled C and D. Following this same labeling structure, this supplement to NUREG-1609 contains two additional appendices, labeled E and F: Appendix E contains basic information on the physical and chemical properties of tritium; Appendix F contains basic information on tritium health physics.

The subsection numbering structure within each section in this supplement to NUREG-1609 is also the same. The fifth subsection is labeled *Review Procedures*, and lists different review approaches for any given subsection. These different review approaches in each Review Procedures subsection are a consequence of significant differences between considerations for the shipment of irradiated TPBARs, and the shipment of any of the other contents described previously in either the parent document, i.e., NUREG-1609, or the MOX supplement to NUREG-1609. Differences that potentially affect the compliance corresponding to the section of the Safety Analysis Report (SAR) in question with NRC regulations will be clearly noted. If no significant differences exist for a particular subsection, that particular subsection is omitted from this supplement to NUREG-1609. Because it is already assumed that the shipment of irradiated TPBARs will be made in packages previously used for the shipment of spent nuclear fuel, numerous cross-references will also be made to individual subsections of NUREG-1617.

Nothing contained in this document may be construed as having the force and effect of NRC regulations (except where the regulations are cited), or as indicating that applications supported by safety analyses and prepared in accordance with RG 7.9 will necessarily be approved, or as relieving any person from the requirements of 10 CFR Parts 20, 30, 40, 60, 70, or 71 or any other pertinent regulations. The principal purpose of this supplement to NUREG-1609 is to ensure the quality and uniformity of staff reviews of packagings intended for transport of irradiated TPBARs. It is also the intent of this plan to make information about regulatory matters widely available, and to improve communications between NRC, interested members of the public, and the nuclear industry, thereby increasing the understanding of the NRC staff review process. In particular, this supplemental guidance, together with NUREG-1609, NUREG-1617, and their previously issued supplements,[4,5] is intended to assist potential applicants by indicating one or more acceptable means of demonstrating compliance with the applicable regulations.

References

1. U.S. Nuclear Regulatory Commission, *Standard Review Plan for Transportation Packages for Radioactive Material*, NUREG-1609, U.S. Government Printing Office, Washington, D.C., 1999.

2. U.S. Nuclear Regulatory Commission, *Standard Format and Content of Part 71 Applications for Approval of Packaging for Radioactive Material*, Regulatory Guide 7.9, Rev. 1, 1986.

3. U.S. Nuclear Regulatory Commission, *Standard Review Plan for Transportation Packages for Spent Nuclear Fuel*, NUREG-1617, U.S. Government Printing Office, Washington, D.C., 1999.

4. U.S. Nuclear Regulatory Commission, *Standard Review Plan for Transportation Packages for MOX-Radioactive Material*, Supplement to NUREG-1609, U.S. Government Printing Office, Washington, D.C., 2002.

5. U.S. Nuclear Regulatory Commission, *Standard Review Plan for Transportation Packages for MOX Spent Nuclear Fuel*, Supplement to NUREG-1617, U.S. Government Printing Office, Washington, D.C., 2002.

1.0   GENERAL INFORMATION REVIEW

1.5   Review Procedures

The general information review of NUREG-1609 would normally be applicable to the review of any packaging used for the shipment of irradiated Tritium-Producing Burnable Absorber Rods (TPBARs). For purposes of this report, however, no specific packaging has been identified for the shipment of such contents. This report, therefore, should be considered to be a topical report, as opposed to a package-specific report.

It is assumed that the packaging to be used will be an existing, modified, or newly designed spent-fuel shipping package. However, because the contents of the package will contain no fissile material, the review format will follow that specified in NUREG-1609.

This section considers each of the subsections of Section 1.5 (Review Procedures) of NUREG-1609 and highlights the special considerations or attention needed for TPBAR shipping packages. In subsections where no significant differences were found, that particular subsection has been omitted from this section.

For all packages, the general information review is based in part on the descriptions and evaluations presented in the Structural Evaluation, Thermal Evaluation, Containment Evaluation, Shielding Evaluation, Criticality Evaluation, Operating Procedures, and Acceptance Tests and Maintenance Program sections of the SAR. Similarly, the results of the general information review are considered in the review of the SAR sections on Structural Evaluation, Thermal Evaluation, Containment Evaluation, Shielding Evaluation, Criticality Evaluation, Operating Procedures, and Acceptance Tests and Maintenance Program.

1.5.2.3 Contents

TPBARs are similar in size and nuclear characteristics to standard, commercial PWR, stainless-steel-clad burnable absorber rods. The exterior of the TPBAR is a stainless-steel tube, approximately 152 inches from tip to tip at room temperature. The nominal outer diameter of the stainless-steel cladding is 0.381 inches. The internal components have been designed and selected to produce and retain tritium.[1-1]

Figure 1-1 illustrates the concentric, cylindrical, internal components of a TPBAR. Within the stainless-steel cladding is a metal getter tube that encircles a stack of annular, ceramic pellets of lithium aluminate. The pellets are enriched with the ^{6}Li isotope. When irradiated in a PWR, the ^{6}Li pellets absorb neutrons, simulating the nuclear characteristics of a burnable absorber rod, and produce tritium, a hydrogen isotope. The tritium chemically reacts with the metal getter, which captures the tritium as a metal hydride.

To meet design limitations on rod internal pressure and burn-up of the lithium pellets, the amount of tritium production per TPBAR is limited to a maximum of 1.2 grams (at 9,619 curies of tritium per gram—see Appendix E) over the full design life of the rod (approximately 500 equivalent full-power days). The potential release rate of tritium into the reactor coolant is subject to a design limit of less than 1,000 Ci/1,000 TPBARs per year. This is achieved by the combined effects of the metal getter tube surrounding the lithium aluminate pellets and an aluminide barrier coating on the inner surface of the cladding.

1.5.2.3.1   TPBAR Components

The TPBAR cladding is double-vacuum-melted, Type 316 stainless steel. To prevent hydrogen from diffusing inward from the coolant to the TPBAR getter and to prevent tritium from diffusing outward from the TPBAR to the reactor coolant, an aluminide coating is on the inner surface of the cladding. This coating barrier must remain effective during fabrication, handling, and in-reactor operations.

The annular ceramic pellets are composed of sintered, high-density, lithium aluminate (LiAlO$_2$).

3

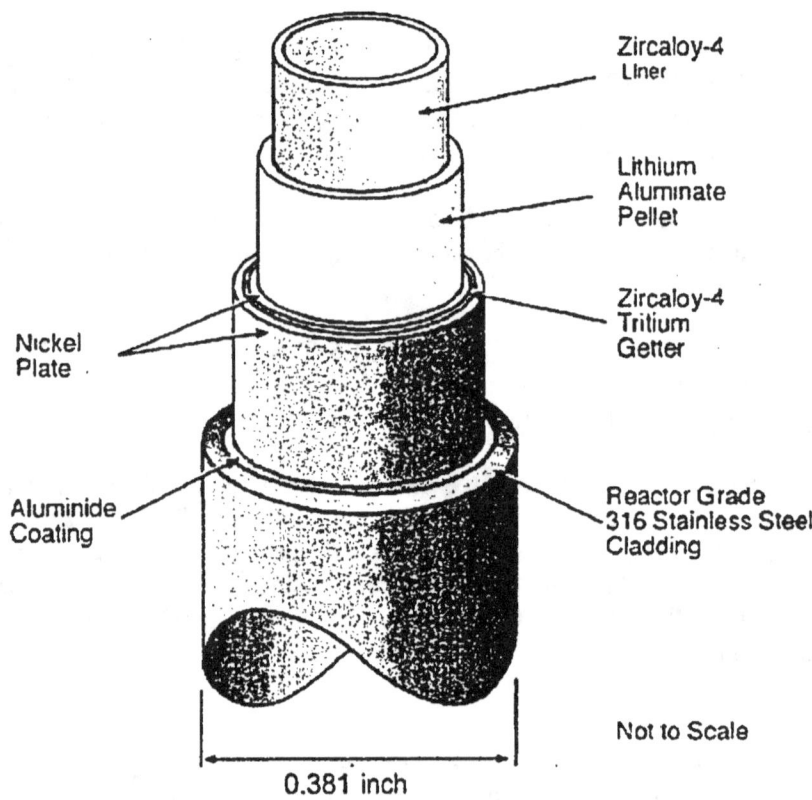

Figure 1-1.   Isometric Section of a Tritium-Producing Burnable Absorber Rod.

The metal getter tube located between the cladding and the lithium aluminate pellets is composed of nickel-plated Zircaloy-4. The getter absorbs the molecular tritium (T_2) generated during irradiation. Nickel plating is used on both sides of the getter to prevent oxidation of the Zircaloy-4 surfaces, which would reduce the tritium absorption rate. Consequently, this plating must remain effective during fabrication, handling, and in-reactor operations.

An unplated Zircaloy-4 tube lines the inside of the annular pellets. This component is called the "liner." Because some of the tritium produced in the pellets may be released as oxidized molecules (T_2O), the liner reduces these species to molecular tritium by reacting with the oxygen. The liner also provides mechanical support to prevent axial movement of pellet material in case any pellets crack during TPBAR handling or operation.

1.5.2.3.2   Axial Arrangement of the Components

Two TPBAR designs are described in this document: 1) the standard TPBAR design, in which the pellet column and getter tubes are segmented into sections called "pencils"; and 2) the full-length getter TPBAR design, in which the getter tube runs the full length of the TPBAR. An "interim option" for the full-length getter design facilitates use of existing pellet stacks and liners.

1.5.2.3.2.1 Standard TPBAR Design

The getter tube is cut and rolled over (coined) to capture the liner and pellets within an assembly called a "pencil." A total of 11 pencil assemblies are stacked within the cladding tube of each TPBAR (see Figure 1-2). The majority of the pencils are of standard length (approximately 12 inches). One or more of the pencils are of variable length.

4

To minimize the impact of power peaking in adjacent fuel rods resulting from the axial gaps between the stacked pencils, there is more than one type of TPBAR. The types are differentiated by where the variable-length pencil or pencils are loaded within the pencil stack. The loading sequence of the pencils is tracked, and each TPBAR is identified by type so that the location of each TPBAR type within a TPBAR assembly can be specified.

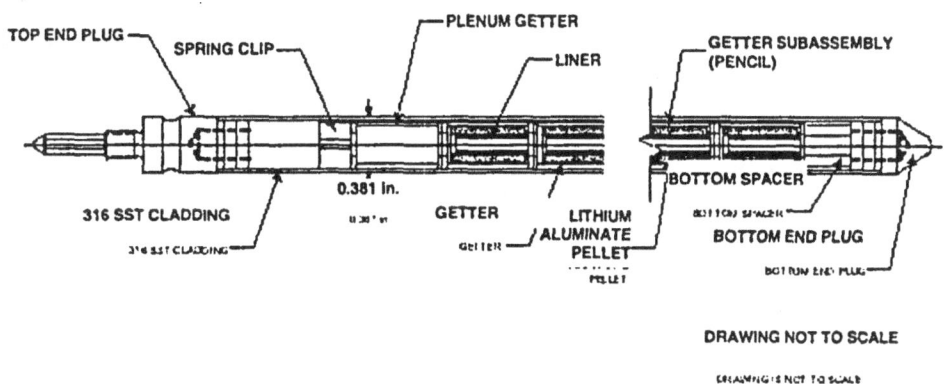

Figure 1-2. Axial Layout of TPBAR Internal Components—Standard Design.

1.5.2.3.2.2 Full-Length Getter TPBAR Design

The axial arrangement of components is altered for the full-length getter TPBAR design. In this design, a single getter tube runs the full length of the TPBAR, and surrounds both the pellet column and the upper and lower spacer tubes (see Figure 1-3). The spacer tubes at the top and bottom of the pellet column are nickel-plated Zircaloy getters. The Zircaloy liner tubes and lithium aluminate pellet stacks in the full-length getter design are longer than in the standard design: typically approximately 16 inches compared to approximately 12 inches in the standard design. However, for the interim full-length getter design option, the liner tubes and pellet stacks will be similar to (or made from) standard-design liner tubes and pellet stacks. That is, a combination of standard length stacks (approximately 12 inches) and short length stacks (approximately 9 inches) from the standard design will be used to make up the pellet column in the interim full-length getter design. The interim design option is employed solely for the purpose of utilizing existing inventories of components.

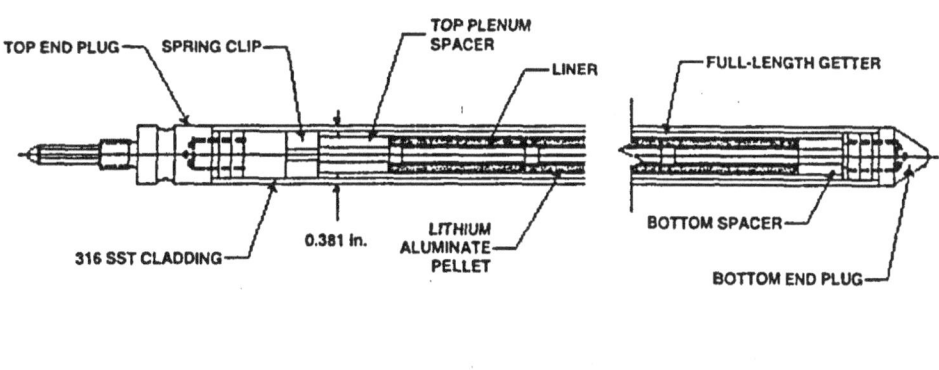

Figure 1-3. Axial Layout of TPBAR Internal Components—Full-Length Getter Design.

The use of the full-length getter design eliminates the need for variable-length pencils and different TPBAR types to minimize the impact of power peaking in adjacent fuel rods resulting from axial gaps between pencils. The

5

pellet column in the full-length getter TPBAR design is essentially continuous, and there is no power peaking penalty from axial gaps in the absorber column.

1.5.2.3.2.3 Common TPBAR Design Features

For hermetic closure of the TPBARs, end plugs similar to those used in commercial PWR burnable absorber rods are welded to each end of the cladding tube. As is shown in Figure 1-2 and Figure 1-3, a gas plenum space is located above the top of the absorber column and below the top end plug. A spring clip in this plenum space holds the internals in place during pre-irradiation handling and shipping. Depending on the design, either a top plenum getter tube or a spacer tube is placed in the plenum space to getter* additional tritium.

The length of the column of enriched lithium aluminate must be variable to provide optimal flexibility in reactor core design. Consequently, the column of enriched lithium aluminate pellets is approximately centered axially about the core mid-plane elevation, but ranges in total length from about 126 to 132 inches. A thick-walled, nickel-plated, Zircaloy-4 spacer tube is placed between the bottom of the absorber column and the bottom end plug both to support the absorber column and to getter tritium.

A TPBAR assembly is shown in Figure 1-4. It should be noted, however, that a typical design used in a 17×17 fuel assembly would be 24 TPBARs, rather than the eight illustrated in Figure 1-4. Multiple fuel assembly designs can be accommodated by changes to the TPBAR lengths and end plugs.

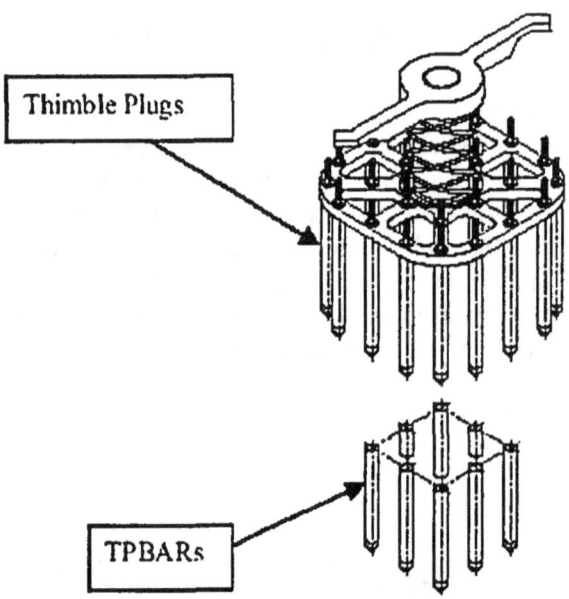

Thimble Plugs

TPBARs

Figure 1-4.   Typical TPBAR Assembly.

After irradiation and removal from the reactor core, the individual TPBARs will be removed from their base plates, and loaded into a consolidation canister for shipment. The consolidation canister, which is designed to hold up to 300 individual TPBARs in a closely packed formation, is then loaded into the shipping container for shipment.

---

*   A colloquial term used in the tritium business, the term *getter* can (and is) often used as a noun, an adjective, and a verb.

Under the current design, therefore, the maximum tritium contents for any given shipment becomes (300 TPBARs) × (1.2 grams of tritium/TPBAR) × (9,619 curies/gram of tritium) ≈ 3.46×10^6 Ci, or about 3,200 A_2. Under these criteria, the package used for the shipment of irradiated TPBARs will be designated as a Category I Package, in accordance with Regulatory Guide 7-11.[1-2]

Other radioactive contents that should be expected include activation products from the stainless-steel cladding. Although these can be expected to include a relatively large fraction of ^{60}Co, the total activity contribution from ^{60}Co should be relatively small, compared to the tritium. The shielding requirements needed for the shipment of irradiated TPBARs, however, are based entirely on the activation products from the stainless steel, and are not driven at all by the tritium.

No fissile material contents are associated with the shipment of irradiated TPBARs. There are, therefore, no criticality concerns.

1.7   References

1-1.   Pacific Northwest National Laboratory, Tritium Technology Program, *Description of the Tritium-Producing Burnable Absorber Rod for the Commercial Light Water Reactor*, TTQP-1-015, Revision 13, August 30, 2004. (Note: The bulk of the material presented in the sections above was taken from this reference.)

1-2.   U.S. Nuclear Regulatory Commission, Regulatory Guide 7-11, *Fracture Toughness Criteria of Base Material for Ferritic Steel Shipping Cask Containment Vessels with a Maximum Wall Thickness of 4 Inches (0.1 m)*, June 1991.

2.0   STRUCTURAL REVIEW

2.5   Review Procedures

The structural review section of NUREG-1609 would normally be applicable to the review of any packaging used for the shipment of irradiated Tritium-Producing Burnable Absorber Rods (TPBARs). For purposes of this report, however, no specific packaging has been identified for the shipment of such contents. This report, therefore, should be considered to be a topical report, as opposed to a package-specific report.

It is assumed that the packaging to be used will be an existing, modified, or newly designed spent-fuel shipping package. However, because the contents of the package will contain no fissile material, the review format will follow that specified in NUREG-1609.

This section considers each of the subsections of Section 2.5 (Review Procedures) of NUREG-1609 and highlights the special considerations or attention needed for TPBAR shipping packages. In subsections where no significant differences were found, that particular subsection has been omitted from this section. A similar situation also pertains to Sections X.5.1, X.5.2.4, and X.5.3.1 of the Spent Fuel Project Office's Interim Staff Guidance document, ISG-15,[2-1] i.e., where no significant differences were found, that particular subsection has been omitted from this section.

For all packages, the structural review is based in part on the descriptions and evaluations presented in the General Information, Thermal Evaluation, Containment Evaluation, Shielding Evaluation, Criticality Evaluation, Operating Procedures, and Acceptance Tests and Maintenance Program sections of the SAR. Similarly, the results of the structural review are considered in the review of the SAR sections on General Information, Thermal Evaluation, Containment Evaluation, Shielding Evaluation, Criticality Evaluation, Operating Procedures, and Acceptance Tests and Maintenance Program.

2.5.2   Materials

2.5.2.1 Material Properties and Specifications

Verify that the effects of tritium, as hydrogen, and helium from the decay of tritium,[†] on the mechanical properties of the structural, bolting, and seal materials have been appropriately taken into consideration, given the assumption that tritium will be released from the TPBARs. (See below; see also Section 4.5.3.)

For containment and other components or materials that may be exposed to tritium, the compatibility of the materials with tritium must be evaluated. Tritium can adversely affect the structural integrity of a material directly or indirectly through a third material. An example of a direct effect is the embrittlement (decrease of ductility or elongation, increase of yield strength, etc.) of a material by tritium dissolved or diffused into the material. High-strength steels are especially susceptible to this embrittlement effect. An example of indirect effect is described in Appendix E: One experiment showed that tritium leached fluorides out of Teflon™ shavings, which subsequently caused stress-corrosion cracking of 316 stainless steel, at high pressures. It is also worth noting that such effects can be highly dependent on both temperature and pressure, and are usually greater

---

[†]   As an isotope of hydrogen, exposure to tritium can be expected to lead to potential hydrogen embrittlement problems for materials that would normally be susceptible to hydrogen embrittlement. The solubility of tritium, however, can also lead to a phenomenon known as *helium embrittlement*, a phenomenon that occurs when tritium finds its way into the material and decays to helium-3. The helium produced by decay gradually migrates to the grain boundaries of the material in question, leading to localized pressure build-ups as a result of the growth of helium bubbles at the grain boundaries. From a materials perspective, therefore, the effects of "...tritium as hydrogen and helium from tritium decay..." are referred to as two different phenomena, and both phenomena must be considered separately. (See also Section E.7.)

at higher temperatures and pressures. Temperature and pressure effects notwithstanding, however, it must also be noted that such effects can be exacerbated greatly in the presence of moisture.

Unfortunately, data concerning tritium effects on shipping containers are rather limited. The package designer is, therefore, obligated to provide a reasonable and conservative estimate of the tritium environment to which each packaging component may be exposed, and a realistic assessment of the potential effects that the tritium environment can have on the properties and structural integrity of each component. The structural reviewer can then determine the significance of the tritium effects to the safety performance of the package. Among all packaging components, those that reside inside, or in close proximity to, the containment boundary have a high risk of tritium effects. Therefore, the relation between the tritium contents and the materials of containment shells, welds, closure bolts, seals, etc., should be thoroughly investigated and understood.

For high-purity tritium containment systems, high-pressure tritium containment systems, and systems where the internal surfaces will be exposed to such environments, 300-series stainless steels are preferred over all other steels. It should also be noted that, for welded assemblies, it is advisable to use only the low-carbon grade (e.g., 304L, 316L, etc.) to reduce the potential for intergranular-corrosion or intergranular-stress-corrosion cracking.

For the shipment of irradiated TPBARs, however, where the internal surfaces of the containment vessel are not expected to see high-purity, or high-pressure tritium environments, the use of other types of stainless steels is acceptable, (1) as long as the material in question has the appropriate structural properties, (2) as long as the material in question is an accepted ASME B&PV Code, Section III material, and (3) as long as additional inspection requirements are imposed, as part of the maintenance program requirements, to guard against long-term problems, such as intergranular-corrosion or intergranular-stress-corrosion cracking. Additional consideration could also be given to limiting the number of times any given package could be used for the shipment of TPBARs. At this point in time, however, no data exist to support such a requirement, and the only way to get these data is through the additional measurements described in Section 7.5.1.2.3, and the additional inspection requirements noted in Section 8.5.2.2. These additional inspection requirements will be needed for all containment components/materials that are reused for multiple TPBAR shipments.

While it may not be possible to predict the actual amount of tritium that may be released into the containment vessel for any given shipment, the information presented in Section 4.5.3 shows that the design criteria for intact TPBARs is <0.12 mCi/(TPBAR·hr), at temperatures between 200° and 650° F. In addition, the information presented in Section 3 shows that the equilibrium temperature for TPBARs during shipment should be about 400° F. From this, it can be seen that, at a minimum, it should be expected that ~300 curies of tritium will be released into the containment vessel, on an annual basis, as a result of normal permeation losses from intact TPBARs. It should also be expected that some number (one or two) of TPBARs *Pre-Failed in Reactor*[‡] could be included in each shipment, for an additional estimate of up to 11.5×10^3 Ci/TPBAR. (See Section 4.5.3.) At a minimum, therefore, it should be assumed that something on the order of 500 curies of tritium will be released into the containment vessel, on an annual basis, for any given shipment. (See also Sections 2.5.6 and 2.5.7, below.) This does not include the additional assumption of the total failure of one (or more) TPBAR(s), with the loss of up to 100% of inventory per TPBAR. (See Table 4-1, and Section 4.5.3.2.2, respectively.)

Using an equilibrium temperature of 400° F, the structural reviewer can begin to make an estimate of the potential effects that a tritium environment can have on the material properties and the structural integrity of each of the containment vessel components. Caution should be exercised, however, for, as was noted above, no actual data exist to support such a conclusion, and the only way to get the actual data is through the additional measurements described in Section 7.5.1.2.3 and the additional inspection requirements noted in Section 8.5.2.2.

[‡]   For a more complete description on TPBARs *Pre-Failed in Reactor*, see the discussion in Section 4.5.3.2.2.

Verify information concerning the accumulation of tritium effects on the materials. Previous exposures to tritium can also affect the repair quality of the affected component. It should be expected that repeated tritium exposures will change the weldability of steels and, thus, the quality of any weld repairs.

2.5.2.2 Prevention of Chemical, Galvanic, or Other Reactions

An overview of a variety of reactions that tritium can have with various materials is provided in Appendix E. All potential reactions, not limited to those affecting only structural properties, should be evaluated, and their possible effects on the safety performance of the package should be assessed. The reviewer should verify that these reactions with tritium, as hydrogen, and helium from the decay of tritium, and their effects on the structural, bolting, and seal materials have been appropriately considered.

The reviewer should also verify that the materials that constitute the TPBARs (i.e., lithium aluminate, Zircaloy-4, etc.) will not have any deleterious chemical, galvanic, and/or other reactions with the containment vessel materials, if the TPBARs are damaged during transportation and storage periods. Because the shipping container is to be loaded under water, and because vacuum-drying processes are to be used prior to shipment (see Section 7.5.1), the presence of moisture should be included in all such considerations.

2.5.2.3 Effects of Radiation on Materials

The reviewer should verify that the damaging effects of radiation from the expected tritium releases from the TPBARs on the structural, bolting, and seal materials have been appropriately considered. Similar to other radioactive materials, tritium can cause degradation or disintegration of plastic materials through radiolysis reactions. (See Appendix E.) However, due to its excellent ability to penetrate materials, tritium can be far more insidious than other radioactive materials. The common practice, as described in Section 4.5.1.1 and in Appendix E, of avoiding the use of elastomeric seals for tritium shipping containers is a direct result of such considerations.

2.5.3 Fabrication and Examination

The reviewer should verify that the effects of tritium, as hydrogen, and helium from the decay of tritium, on the fabrication procedures and examination requirements of the containment system have been appropriately considered, assuming that tritium will be released from the irradiated TPBARs.

As noted in Section 2.5.2.1, components or materials that have been previously exposed to tritium may need special repair procedures and/or post-repair examinations.

Special precautions should be taken to control and qualify weld materials, weld processes, weld procedures, and welders, as appropriate, for the materials selected for the containment body and lid. Additional precautions should also be taken to note that the appropriate follow-up procedures have been added to long-term maintenance requirements for the packaging, again, to guard against long-term problems such as intergranular-corrosion or intergranular-stress-corrosion cracking. See Table 2 of Reference 2-2 for a summary of welding criteria that are based on the requirements of the ASME B&PV Code.

2.5.4 Lifting and Tie-Down Standards for All Packages

The lifting and tie-down devices of a TPBAR shipping package should not normally be exposed to tritium. Therefore, the evaluation of such devices should be no different for a TPBAR shipping package than for other packages. However, if such devices are an integral part of the containment vessel, such as trunnions attached to the containment vessel, the reviewer should verify that the structural capacity of the trunnions will not be degraded by tritium that may have permeated through the containment vessel after multiple shipments.

2.5.6 Structural Evaluation under Normal Conditions of Transport

The reviewer should verify that the structural, bolting, and seal components/materials can uphold the safety performance of the package under Normal Conditions of Transport, if the components have been exposed to and may be affected by contact with tritium.

As discussed in Section 4.5.1.1.3, elastomeric seals cannot used for the containment of tritium. The containment seals of tritium packages are commonly made of metal O-rings or metal-to-metal, knife-edge seals. These types of seals typically require a greater compression than that needed for elastomeric seals. To provide the necessary compression, high strength bolts are often used with a high preload. The high preload is also intended to prevent vibrational loosening of the bolted closure, which can occur during Normal Conditions of Transport. Using a very high preload (sometimes as much as 90% of the proof load of the bolts) is a common practice for preventing vibrational loosening. However, because high-strength bolts are susceptible to embrittlement by tritium, the high preload may cause the bolts to fracture unexpectedly under Cold Conditions, if the bolts have been affected by tritium. Normally, the fracture of a single bolt should not result in the fracture of other bolts and a catastrophic failure of the containment closure. Thus, Regulatory Guides (Reg. Guides) 7.11 and 7.12 have not explicitly included the containment closure bolts as "fracture critical" components, whose fracture, once initiated, will continue, and result in a catastrophic failure of the containment.[2-3, 2-4] Thus, closure bolts of most packages are exempt from the stringent fracture-toughness requirement specified in Reg. Guides 7.11 and 7.12. However, in the case of tritium containment, with high-strength bolts and high bolt preloads, such an exemption may not be a prudent practice. Therefore, it is recommended that the fracture criteria of Reg. Guides 7.11 and 7.12 also be used for the selection of closure bolts for TPBAR shipping packages. In addition, the bolt stress should be kept below the bolting stress limits of ASME B&PV Code, Section III, Subsection NB. Thus, methods other than using very high preload may be needed to prevent vibrational loosening.

As discussed above in Section 2.5.2.1, the package designer is obligated to provide a reasonable and conservative estimate of the tritium environment to which each packaging component may be exposed, and a realistic assessment of the potential effects that the tritium environment can have on the properties and structural integrity of each component. As indicated in Table 4-1, the amount of tritium released from damaged TPBARs can be several orders of magnitude greater than that from intact TPBARs, or from event-failed TPBARs. Thus, the tritium concentration within the containment boundary can increase significantly with an increasing number of damaged TPBARs. For Normal Conditions of Transport, the condition that has the greatest potential to produce additional damage to the TPBARs is vibration. A vibration and fatigue evaluation of the TPBARs should be performed to determine if the natural frequencies of the TPBARs lie in the dominant frequency ranges of the transport vehicle floor. While there are no regulatory requirements that state that the contents must arrive at the destination site intact, it is important to note that the working lifetimes of the components exposed to tritium can be expected to be inversely proportional to the tritium levels to which the components are exposed.

2.5.7 Structural Evaluation under Hypothetical Accident Conditions

The reviewer should verify that excessive damage of the irradiated TPBAR contents will not occur under Hypothetical Accident Conditions, so that the safety performance of the package will not be catastrophically affected throughout the sequence of Hypothetical Accident Condition tests.

As was noted above, the amount of tritium released from damaged TPBARs can be several orders of magnitude greater than that from intact TPBARs, or from event-failed TPBARs, and that the tritium concentration in the containment can increase significantly with an increasing number of damaged TPBARs. Under Hypothetical Accident Conditions, the test requirement that can be expected to have the greatest potential to produce damage to the TPBARs is the 30-ft end-on drop. A buckling analysis of the TPBARs should, therefore, be performed for the 30-ft end-on drop. Under the large axial compression generated by the end-on drop, the long, slender TPBARs can buckle easily and rupture after suffering excessive deformation/strain after buckling. The buckling evaluation of TPBARs must employ realistic assumptions about the initial geometric imperfections, as well as the lateral and end constraints of the TPBARs. When the effects of geometric imperfections and constraints are properly included, it should be expected that inadequately supported TPBARs can buckle easily under relatively low impact g loads. The reviewer, therefore, should verify that the TPBARs will be properly supported throughout the entire sequence of Hypothetical Accident Condition tests.

Again, as was noted above, there are no regulatory requirements that state that the contents must arrive at the destination site intact. In this case, however, the reviewer should be looking for the possibility of catastrophic failure of the containment vessel, or any of its major components, as a result of substantially increased levels of tritium into containment.

2.7   References

2-1.   U.S. Nuclear Regulatory Commission, Spent Fuel Project Office, *Materials Evaluation*, Interim Staff Guidance-15, January 10, 2001.

2-2.   Monroe, R.E., Woo, H.H., and Sears, R.G., Lawrence Livermore National Laboratory, *Recommended Welding Criteria for Use in the Fabrication of Shipping Containers for Radioactive Materials*, NUREG/CR-3019, U.S. Nuclear Regulatory Commission, 1984.

2-3.   U.S. Nuclear Regulatory Commission, Regulatory Guide 7-11, *Fracture Toughness Criteria of Base Material for Ferritic Steel Shipping Cask Containment Vessels with a Maximum Wall Thickness of 4 Inches (0.1 m)*, June 1991.

2-4.   U.S. Nuclear Regulatory Commission, Regulatory Guide 7-12, *Fracture Toughness Criteria of Base Material for Ferritic Steel Shipping Cask Containment Vessels with a Wall Thickness Greater than 4 Inches (0.1 m) But Not Exceeding 12 Inches (0.3 m)*, June 1991.

3.0   THERMAL REVIEW

3.5   Review Procedures

The thermal review section of NUREG-1609 would normally be applicable to the review of any packaging used for the shipment of irradiated Tritium-Producing Burnable Absorber Rods (TPBARs). For purposes of this report, however, no specific packaging has been identified for the shipment of such contents. This report, therefore, should be considered to be a topical report, as opposed to a package-specific report.

It is assumed that the packaging to be used will be an existing, modified, or newly designed spent-fuel shipping package. However, because the contents of the package will contain no fissile material, the review format will follow that specified in NUREG-1609.

This section considers each of the subsections of Section 3.5 (Review Procedures) of NUREG-1609 and highlights the special considerations or attention needed for TPBAR shipping packages. In subsections where no significant differences were found, that particular subsection has been omitted from this section.

For all packages, the thermal review is based in part on the descriptions and evaluations presented in the General Information, Structural Evaluation, Containment Evaluation, Shielding Evaluation, Criticality Evaluation, Operating Procedures, and Acceptance Tests and Maintenance Program sections of the SAR. Similarly, the results of the thermal review are considered in the review of the SAR sections on General Information, Structural Evaluation, Containment Evaluation, Shielding Evaluation, Criticality Evaluation, Operating Procedures, and Acceptance Tests and Maintenance Program.

3.5.1   Description of Thermal Design

3.5.1.2   Content Decay Heat

According to Table 4 from Reference 3-1 (reproduced below as Table 3-1), the TPBAR heat load 30 days after removal from the reactor is estimated by the design agency to be 3.35 W/TPBAR. Although the estimated value quickly drops to 2.31 W/TPBAR at a 90-day time interval, for purposes of conservatism, the 30-day value should be used for all thermal analyses, throughout.

This is also consistent with the information presented in Section 2.10.6 of Reference 3-2, which states that,

> "TVA [has] also evaluated the heat production from a fully loaded consolidation canister and its potential effect on the spent fuel racks. The potential heat generation within the consolidation canister is small enough that it can be safely stored in the existing fuel racks. An irradiated absorber rod will only produce about 3 watts of heat 30 days after reactor shutdown. This is equivalent to a maximum heat load of 900 watts/canister, assuming a fully loaded canister contains a maximum of 300 absorber rods. This heat load is small given that adequate circulation is provided through the open topped canister and through the drainage/cooling holes on the sides and bottom of the canisters. Therefore, the staff concludes that this configuration will provide adequate natural circulation."[3-2]

Since the typical heat load for a spent-fuel shipping container is normally on the order of a few- to several-hundred kW, the total heat load on a typical TPBAR transport package should be relatively small. In the case of a TPBAR transport package, however, the total heat load is not particularly important. What is more important is the equilibrium temperature of the consolidated bundle of TPBARS within the containment vessel, since temperature will be the primary driving force for the expected tritium losses from the TPBARs into containment. Preliminary analyses suggest that the equilibrium temperature should be on the order of ~400° F. (See the related discussions in Section 2.5.2.1, above, and Sections 3.5.4.2 and 4.5.3, below.)

Table 3-1. Decay Heat in a TPBAR (Watts/TPBAR)

Nuclide	7 Days	30 Days	90 Days	180 Days	1 Year	5 Years	10 Years
^{3}H§	3.90E-01	3.89E-01	3.85E-01	3.80E-01	3.69E-01	2.95E-01	2.23E-01
^{32}P	1.04E-02	3.42E-03	1.87E-04	2.38E-06	3.06E-10	5.86E-12	5.83E-12
^{51}Cr	2.07E-01	1.17E-01	2.60E-02	2.74E-03	2.66E-05	3.57E-21	5.10E-41
^{54}Mn	2.09E-01	1.98E-01	1.73E-01	1.42E-01	9.42E-02	3.69E-03	6.42E-05
^{55}Fe	7.28E-03	7.15E-03	6.85E-03	6.41E-03	5.60E-03	1.93E-03	5.08E-04
^{59}Fe	1.54E-01	1.08E-01	4.28E-02	1.07E-02	6.16E-04	1.04E-13	6.30E-26
^{58}Co	1.61E+00	1.29E+00	7.14E-01	2.96E-01	4.82E-02	2.94E-08	5.03E-16
^{60}Co	5.55E-01	5.50E-01	5.39E-01	5.21E-01	4.88E-01	2.88E-01	1.49E-01
^{63}Ni	2.30E-03	2.30E-03	2.30E-03	2.30E-03	2.29E-03	2.22E-03	2.14E-03
^{76}As	7.74E-03	3.76E-09	1.28E-25	0.00E+00	0.00E+00	0.00E+00	0.00E+00
^{95}Zr	3.33E-01	2.60E-01	1.36E-01	5.11E-02	6.87E-03	9.18E-10	2.35E-18
^{95}Nb	3.32E-01	3.12E-01	2.13E-01	9.53E-02	1.41E-02	1.93E-09	4.93E-18
^{99}Mo	5.40E-02	1.64E-04	4.44E-11	6.24E-21	0.00E+00	0.00E+00	0.00E+00
^{117m}Sn	1.52E-02	4.88E-03	2.50E-04	2.91E-06	3.03E-10	0.00E+00	0.00E+00
^{119m}Sn	4.35E-03	4.08E-03	3.44E-03	2.67E-03	1.58E-03	2.53E-05	1.45E-07
^{125}Sn	1.46E-02	2.79E-03	3.73E-05	5.77E-08	9.47E-14	0.00E+00	0.00E+00
^{125}Sb	5.23E-03	5.20E-03	5.00E-03	4.70E-03	4.14E-03	1.52E-03	4.35E-04
^{182}Ta	9.55E-02	8.31E-02	5.79E-02	3.36E-02	1.10E-02	1.65E-06	3.42E-11
^{183}Ta	1.61E-01	7.08E-03	2.03E-06	9.91E-12	1.15E-22	0.00E+00	0.00E+00
Total	4.19E+00	3.35E+00	2.31E+00	1.55E+00	1.05E+00	5.92E-01	3.75E-01

3.5.4 Thermal Evaluation under Normal Conditions of Transport

3.5.4.2 Maximum Normal Operating Pressure

For TPBAR transport packages, the maximum normal operating pressure (MNOP) at the estimated temperature of about 400° F should be in the range of one to two atmospheres, plus any additional pressure generated due to tritium in-leakage/permeation. It should be noted, however, that, based on the information presented in Section 4.5.3, below, the additional pressure generated due to tritium in-leakage/permeation is only expected to range between 7.6×10^{-6} and 5.2×10^{-3} moles of tritium per year, for intact TPBARs (see Table 4-1). As such, the additional pressure generated due to tritium in-leakage/permeation would likely be a second-order correction.

The requirement that tritium (as hydrogen) makes up less then 5% of the gas for flammability regulations is also satisfied because, as is shown above, the contribution of tritium (as hydrogen), as a flammable gas, can be expected to be small. In addition, it should also be noted that, any tritium that escapes from intact TPBARs will be rapidly converted to HTO.** As tritiated water vapor, the available tritium (i.e., as HTO) is already oxidized and, therefore, is no longer flammable. As yet a third layer of conservatism, the reviewer should verify that, as part of the loading process, the package will be vacuum dried and backfilled with an inert gas, in accordance with the generic procedures outlined in the Pacific Northwest National Laboratory document, *Evaluation of Cover Gas Impurities and Their Effects on the Dry Storage of LWR Spent Fuel*.[3-4] This should be verified as part of the Operating Procedures review.

---

§   The ORIGEN2 values for H-3 are not reported. The values given for H-3 are based on a maximum of 1.2 g of tritium per TPBAR at discharge, as specified in Reference 3-3. There is 0.325 W per gram of tritium, and the half-life of tritium is 12.33 years. The value of 1.2 g at discharge is decayed appropriately for the various decay times.

**   Chemically, the term HTO is used to describe tritiated water vapor. (See Appendix E.) While that may be more favorable from a transportation perspective, it is not nearly as favorable from a health and safety perspective because HTO is, by far, more hazardous than tritium gas (i.e., HT or T_2). (See Appendix F.)

16

For those situations where the tritium released into containment might be substantially greater than that described above, such as the total failure of one (or more) TPBARs, with the loss of up to 100% of inventory per TPBAR, the reviewer should verify that the tritium concentration in any void volume of the containment will be less than 5%, by volume, over the standard shipping time of one year.

One additional factor that must be considered is a possible change in the thermal properties of the backfill gas. As a first approximation, it should be assumed that the thermal properties of tritium are virtually identical to those of hydrogen. Likewise, it should also be assumed that the thermal properties of tritiated water vapor (HTO) are virtually identical to those of normal water vapor (H_2O). As long as the tritium losses into containment are small, such as those described above, i.e., between 7.6×10^{-6} and 5.2×10^{-3} moles of tritium per year, changes to the thermal properties of the backfill gas would likely be negligible. As the estimated tritium losses into containment get larger, such as those described below in Section 4.5.3, i.e., on the order of ~0.2 moles of tritium, or more, the reviewer should verify that the applicant has provided the appropriate calculations 1) using the assumption of 100% tritium (as hydrogen) gas, and 2) using the assumption of 100% HTO. The worst-case situation can then be determined, and verified, by the reviewer.

3.5.5 Thermal Evaluation under Hypothetical Accident Conditions

3.5.5.3 Maximum Temperatures and Pressures

As an absolute, worst-case condition, the reviewer should assume that all TPBARs fail, with the loss of up to 100% of the total tritium inventory. This would be equivalent to a total loss of $\sim 3.46 \times 10^6$ curies, or ~ 60 moles of tritium.

As a first approximation, the estimated temperature of the TPBARs and the surrounding gas should be about 400° F.

As for possible changes to the thermal properties of the back-fill gas, the reviewer should again verify that the applicant has provided the appropriate calculations 1) using the assumption of 100% tritium (as hydrogen) gas, and 2) using the assumption of 100% HTO. The worst-case situation can then be determined, and verified, by the reviewer.

3.7 References

3-1.   Pacific Northwest National Laboratory, Tritium Technology Program, *Unclassified Bounding Source Term, Radionuclide Concentrations, Decay Heat, and Dose Rates for the Production TPBAR,* TTQP-1-111, Revision 4, September 16, 2004.

3-2.   U.S. Nuclear Regulatory Commission, *Safety Evaluation by the Office of Nuclear Reactor Regulation Related to Amendment No. 40 to Facility Operating License No. NPF-90 Tennessee Valley Authority Watts Bar Nuclear Plant, Unit 1 Docket No. 50-390,* September 23, 2002. (See, in particular, Section 2.10.6.) Note: This particular document was included as Enclosure 2, as part of a letter, L. M. Padovan (NRC), to J. A. Scalice (TVA), dtd., September 23, 2002, Subject: Watts Bar Nuclear Plant, Unit 1— Issuance of Amendment to Irradiate up to 2,304 Tritium-Producing Burnable Absorber Rods in the Reactor Core (TAC NO. MB1884), ADAMS Accession No. ML022540925.

3-3.   Lopez Jr. A., 2003, *Production TPBAR Design Inputs for Watts Bar (U),* PNNL-TTQP-1-702, Rev. 9., Pacific Northwest National Laboratory, Richland, Washington.

3-4.   Knoll, R.W. and Gilbert, E.R., *Evaluation of Cover Gas Impurities and Their Effects on the Dry Storage of LWR Spent Fuel,* PNL-6365, Pacific Northwest National Laboratory, Richland, Washington, November 1987.

4.0 CONTAINMENT REVIEW

4.5 Review Procedures

The containment review section of NUREG-1609 would normally be applicable to the review of any packaging used for the shipment of irradiated Tritium-Producing Burnable Absorber Rods (TPBARs). For purposes of this report, however, no specific packaging has been identified for the shipment of such contents. This report, therefore, should be considered to be a topical report, as opposed to a package-specific report.

It is assumed that the packaging to be used will be an existing, modified, or newly designed spent-fuel shipping package. However, because the contents of the package will contain no fissile material, the review format will follow that specified in NUREG-1609.

This section considers each of the subsections of Section 4.5 (Review Procedures) of NUREG-1609 and highlights the special considerations or attention needed for TPBAR shipping packages. In subsections where no significant differences were found, that particular subsection has been omitted from this section. A similar situation also pertains to Section X.5.2.9 of the Spent Fuel Project Office's Interim Staff Guidance document, ISG-15,[4-1] i.e., where no significant differences were found, that particular subsection has been omitted from this section.

For all packages, the containment review is based in part on the descriptions and evaluations presented in the General Information, Structural Evaluation, Thermal Evaluation, Shielding Evaluation, Criticality Evaluation, Operating Procedures, and Acceptance Tests and Maintenance Program sections of the SAR. Similarly, the results of the containment review are considered in the review of the SAR sections on General Information, Structural Evaluation, Thermal Evaluation, Shielding Evaluation, Criticality Evaluation, Operating Procedures, and Acceptance Tests and Maintenance Program.

4.5.1 Description of the Containment System

4.5.1.1 Containment Boundary

4.5.1.1.1 Materials of Construction

For high-purity tritium containment systems, high-pressure tritium containment systems, and systems where the internal surfaces will be exposed to such environments, 300-series stainless steels are preferred over virtually all other materials. It should also be noted that, for welded assemblies, it is advisable to use only the low-carbon grades (e.g., 304L, 316L, etc.) to reduce susceptibility to intergranular-corrosion or intergranular-stress-corrosion cracking.

For the shipment of irradiated TPBARs, however, where the internal surfaces of the containment vessel are not expected to see high-purity or high-pressure tritium environments, the use of other types of stainless steels is acceptable, 1) as long as the material in question has the appropriate structural properties, 2) as long as the material in question is an accepted ASME B&PV Code, Section III material, and 3) as long as additional inspection requirements are imposed, as part of the maintenance program requirements, to guard against long-term problems such as intergranular-corrosion or intergranular-stress-corrosion cracking. (See also the related discussions in Section 2.5.2.1, above, and Section 8.5.2.2, below.)

4.5.1.1.2 Welds

Special precautions should be taken to control and qualify weld materials, weld processes, welding procedures, and welders, as appropriate, for the material selected for the containment vessel body and lid. Additional precautions should also be taken to note that the appropriate follow-up procedures have been added to long-term maintenance requirements for the packaging, again, to guard against long-term problems such as intergranular-corrosion or intergranular-stress-corrosion cracking. (See Table 2 of Reference 4-2 for a summary of welding criteria that is based on the requirements of the ASME Boiler and Pressure Vessel Code. See also Section 8.5.2.2, below.)

4.5.1.1.3   Seals

The generic rule of thumb for any tritium handling system is that elastomeric seals[tt] are not acceptable for use in any part of the containment boundary. This includes 1) the use of elastomeric seals between the containment vessel body and lid, 2) the use of elastomeric seals for any valve stem tip/valve seat combinations that might be part of the containment boundary, such as vent- and drain-port valves, and/or 3) the use of elastomeric seals between the containment vessel body and the vent- and drain-port covers, when the vent- and drain-port covers are part of the containment boundary. The primary reason for this general prohibition on the use of elastomeric seals can be traced, in part, to permeation issues and, in part, to the requirements of ANSI N14.5-1997.[4-3] As is noted in Section B.11 of ANSI N14.5,

> "Permeation is the passage of a fluid through a solid barrier ... by adsorption-diffusion-desorption processes. It should not be considered as leakage or a release unless the fluid itself is hazardous or radioactive. If this is the case, the container boundary must reduce the permeation to an acceptable level."[4-3]

Since the permeation rate of tritium through most elastomers is about two orders of magnitude higher than that allowed by regulatory limits, the use of elastomeric seals cannot be allowed. (See the additional information presented in Appendix E.)

The use of elastomers/elastomeric seals is also discouraged for valve stem tip/valve seat combinations in those situations where the vent- and drain-port valves might become part of the containment boundary and in any situation where the surface of the elastomer might be wetted with tritium. In this case, however, the general prohibition stems from the chemical and physical properties of tritium, and from the tendency of tritium to form undesirable chemical by-products, which can lead to the long-term degradation of the containment boundary. (See Sections E.7 and E.8.)

The preferred methods for sealing systems that are designed to contain tritium are through the use of all-welded construction. When the use of all-welded construction is not realistic, such as the containment boundary seal areas for transportation packages with bolted closures, the use of metal seals and/or metallic O-rings is recommended.

4.5.1.2   Special Requirements for Plutonium

This requirement is not applicable to the shipment of irradiated TPBARs. It should also be noted that this requirement is no longer part of the requirements for Type B packagings, as per the October 2004 changes to 10 CFR Part 71.

4.5.2   General Considerations

4.5.2.2 Type B Packages

Section 4.5.2.2 of NUREG-1609 specifies that Type B packagings must satisfy the quantified release rates in §71.51 of 10 CFR Part 71. An acceptable method for satisfying these requirements is provided in ANSI N14.5. Additional information for the determination of containment criteria is also provided in NUREG/CR-6487.[4-4] Additional discussion is also provided below in Section 4.5.3.

[tt]   For purposes of this document, the term *elastomeric seals* pertains equally to organic, elastomeric, halogenated hydrocarbon, thermoplastic resin, and/or thermosetting resin types of seals. See Appendix E.

4.5.2.3 Combustible-Gas Generation

As is noted above in Section 3.5.4.2, the bulk of the gases releases from irradiated TPBARs under Normal Conditions of Transport will be released as HTO,[‡‡] or tritiated water vapor. As tritiated water vapor, the available tritium (i.e., as HTO) is already oxidized and, therefore, is no longer flammable. An additional layer of conservatism is added, and the reviewer should verify that, as part of the loading process, the package will be vacuum dried and backfilled with an inert gas, in accordance with the generic procedures outlined in the Pacific Northwest National Laboratory document, *Evaluation of Cover Gas Impurities and Their Effects on the Dry Storage of LWR Spent Fuel.*[4-5] For Normal Conditions of Transport, therefore, with no unexpected TPBAR failures (see below), there should be no possibility for the formation of a combustible-gas mixture inside the containment boundary.

For those situations where the tritium released into containment might be substantially greater than that described above, such as the total failure of one (or more) TPBARs, with the loss of up to 100% of inventory per TPBAR, the reviewer should verify that the tritium concentration in any void volume of the containment will be less than 5%, by volume, over the standard shipping time of one year.

Under Hypothetical Accident Conditions, the situation can change, in that the tritium concentrations, as T_2 or HT, could be relatively high. In this case, however, a monitoring technique is discussed briefly in Section 7.5.1.2.3 that can be used to determine the actual tritium concentration inside containment, which, on an as needed basis, can also be used to determine potential flammability levels of the gases inside containment. Use of this technique is discussed more fully in Chapter 7.

4.5.3   Containment under Normal Conditions of Transport (Type B Packages)

4.5.3.1 Containment Design Criteria

The generic rule of thumb for any package intended for the shipment of tritium is that the package will have to be designed to meet the ANSI N14.5 definition of *leaktight* for Normal Conditions of Transport. By definition, therefore, the allowable leakage criterion specified should be $\leq 1 \times 10^{-7}$ reference·cm^3/s. Also, by definition, the adoption of the leaktight criterion eliminates the applicant's need to justify any containment-boundary design criteria calculations.

On the other hand, the applicant could elect to follow the guidance presented in Chapter 4 of NUREG/CR-6487 for the determination of a source term for dispersible radioactive solids that might be entrained in the tritium that is also available for release. The determination of the source term for the available radioactive solids is relatively straightforward, because the design agency for the TPBARs (Pacific Northwest National Laboratories, or PNNL) has made that information available.[4-6] What complicates the problem, in this case, is the determination of the amount of tritium that might be available.

In a separate supporting document, the design agency for the TPBARs has also provided some estimates for potential release rates of tritium into the containment boundary.[4-7] A summary of these estimates is provided below in Table 4-1. The information therein was adapted from Reference 4-7.

4.5.3.2 Demonstration of Compliance with Containment Design Criterion

A review of these estimates suggests that it would be difficult, if not impossible, to determine an actual source term to be used for the determination of an allowable release rate for a package to be used for the shipment of TPBARs. A review of the information provided by the design agency is worthwhile, however, because the estimates provided here can be used to determine the condition of the TPBARs, after they have been

---

[‡‡]   Chemically, the term HTO is used to describe tritiated water vapor. (See Appendix E.) While that may be more favorable from a transportation perspective, it is not nearly as favorable from a health and safety perspective because HTO is, by far, more hazardous than tritium gas (i.e., HT or T_2). (See Appendix F.)

consolidated,[§§] and after they have been loaded into the containment vessel. (Note: The release estimates cited below in Table 4-1 are the actual design criteria for both, the *Standard* TPBAR design, and the *Full-Length* TPBAR design, respectively; see Section 1.5.2.3.2.)

4.5.3.2.1   TPBAR Containment System Design Criteria, Intact TPBARs

Under the broader heading of Normal Conditions of Transport, the design agency's estimate of <0.05 mCi/hr for 1,200 or fewer TPBARs, in the first column of Table 4-1, is actually not appropriate for use as a source term for the releasable tritium, because the temperature estimates for the TPBARs in a consolidated bundle of up to 300 TPBARs should be more on the order of ~400° F. (See Section 3.5.4.) This is, however, an excellent place to start because it does point out an operational fact that there *will* be permeation losses from the TPBARs, under Normal Conditions of Transport, and that these permeation losses *will* be going directly into containment.

The estimate provided by the design agency of <0.05 mCi/hr for the consolidated contents (i.e., up to 300 TPBARs) further equates to ~8.40 mCi/week and, for MNOP determination timeframes, ~437 mCi/yr, or ~7.6×10^{-6} moles of tritium per year. At the permeation rate cited in this case, all of the tritium would rapidly be converted to HTO, as soon as it is released, and combustible-gas generation issues will not be an issue. (See Section 3.5.4.2, above, and Sections E.5 and E.6, below.)

Table 4-1. Summary of Tritium Release Assumptions for Cask Transportation Scenarios.

Intact TPBARs (Normal Conditions of Transport)		Event-Failed TPBARs (Hypothetical Accident Conditions)		TPBARs Pre-Failed In-Reactor	
<200° F	200° F to 650° F	Ambient to <200° F	200° F to 650° F	Ambient to <200° F	>200° F
<0.05 mCi per hour for 1,200 or fewer TPBARs	<0.12 mCi per TPBAR per hour (based on average TPBAR in the core)	<0.1 Ci per TPBAR per hour, not to exceed 1% of the pellet tritium inventory	<55 Ci total per TPBAR	<0.1 Ci per TPBAR per hour	Up to 100% of inventory

The design agency's estimate of <0.12 mCi/(TPBAR·hr), in the second column of Table 4-1, is not really appropriate either, because it is a simple data reduction value for the reactor in-core estimated permeation releases. The design agency has stated that, for intact TPBARs, "The in-reactor design tritium release rate for TPBARs is less than 1,000 Ci per 1,000 rods per year. The in-reactor design tritium release rate should be used on a core-averaged basis. This release rate should not be applied as a limit for individual TPBARs."[4-7] In additional supporting documentation, further clarification was added to note that,

> "... (T)he TPBARs were designed such that permeation through the cladding would be less than 1.0 Ci/TPBAR/year. For the production design, this value is reported as 'less than 1000 Ci/1000 TPBAR/year.' While the value of the permeation is not changed ..., the new units of reporting emphasize that the release is based on the core average. Thus, while an individual TPBAR may release more than 1 Ci/year, the total release for 1,000 TPBARs will be less than 1,000 Ci/year."[4-8]

[§§] Additional information on *consolidation* and the *pre-shipment* and *post-shipment* measurements is provided in Section 7 of this document.

Although a value of <0.12 mCi/(TPBAR·hr) may not be useful as a source term for transportation purposes, it does serve a useful operational purpose, because, like the estimate provided for the first column of Table 4-1, it does provide a second data point toward the determination of possible tritium permeation losses into containment.

As has already been noted, a value of <0.12 mCi/(TPBAR·hr) translates to ~20.2 mCi/(TPBAR·week) and, for MNOP purposes, to ~1 Ci/(TPBAR·yr). For consolidated shipments of up to 300 TPBARs, this further translates to ~300 Ci/yr, or ~5.2×10^{-3} moles of tritium per year, going into containment. Again, at the permeation rate cited in this case, all of the tritium would rapidly be converted to HTO—see Section 3.5.4.2 and Appendix E—as soon as it is released, so combustible-gas generation issues should not be an issue.

4.5.3.2.2 *TPBAR Containment System Design Criteria, TPBARs Pre-Failed In-Reactor*[***]

For those situations where the tritium released into containment might be substantially greater than that described in either of the situations noted above, such as the total failure of one (or more) TPBARs, two different scenarios are listed in Table 4-1 under the heading of *TPBARs Pre-Failed In-Reactor*: 1) where the temperature estimate is ambient to <200° F, and 2) where the temperature estimate is >200° F. Both situations should be considered under the broader heading of Normal Conditions of Transport. However, because the estimated equilibrium temperature of the TPBARs, under Normal Conditions of Transport, is expected to be closer to 400° F, the >200° F scenario is both bounding, and more realistic, and the ambient to <200° F scenario need not be considered any further.

Under the right-hand-most column in Table 4-1, the potential loss of up to 100% of the inventory per TPBAR represents an addition to the source term that should be used for estimating the total tritium losses into containment for Normal Conditions of Transport. As a bounding value, this represents an additional loss of 1.2 grams, 11,543 Ci, or ~0.20 moles of tritium, per TPBAR, going into containment. Since the possibility that some of the losses may not be fully converted to HTO cannot be ruled out in this case, it should, therefore, be assumed that some of the losses from the TPBAR will be as T_2 and/or HT. The reviewer, therefore, should verify that the combustible-gas (i.e., the tritium) concentration in any void volume of the containment will be less than 5%, by volume, over the standard MNOP shipping time of one year. Such an assessment should include the possibility that one, or more, TPBARs might fail in this manner, for any given shipment.

4.5.4   Containment under Hypothetical Accident Conditions (Type B Packages)

4.5.4.1 Containment Design Criterion

For Hypothetical Accident Conditions (HAC), the applicant is left with the same two options that were available above for Normal Conditions of Transport (NCT): 1) Adopt the *leaktight* criterion, as specified in ANSI N14.5, with no additional calculations or supporting justification, or 2) adopt a bounding calculation, which would include the assumption of a total tritium loss, along with the assumption of the aerosol losses from the activation products, and the applicant would have to justify all calculations for the source term. The primary difference between the two options is that, unlike the situation for NCT, a bounding source term for the available tritium would be relatively easy to define, and the resulting calculations should push the leakage test criteria into the 10^{-5} range, as opposed to that used for leaktight, i.e., $\leq 1 \times 10^{-7}$ reference·cm^3/s. In either case, however, additional input will be required from both the structural and thermal sections of the application to show that there will be no unexpected deformation in the area around the containment seals as a result of the HAC testing requirements, and that the HAC temperature requirements will not compromise containment boundary seals.

[***] By definition, the term, *Pre-Failed In-Reactor*, is intended to address the possibility of a TPBAR weld failure that occurs just before the TPBARs are unloaded from the reactor core. An NCT situation, this scenario further assumes that the TPBAR in question becomes water logged, prior to being consolidated with the other TPBARs, and prior to being loaded into the shipping container. Between the chemical reactions that would be expected to occur between the water and the internal components of the TPBAR, and the expected increase in temperature, the TPBAR(s) in question would be expected to lose up to 100% of its (their) inventory. (See Reference 4-7.)

4.5.4.1.1   TPBAR Containment System Design Criteria, Event-Failed TPBARs[†††]

Two different scenarios are listed in Table 4-1 under the heading of *Event-Failed TPBARs*: 1) where the temperature estimate is ambient to <200° F, and 2) where the temperature estimate is >200° F. Both situations should be considered under the broader heading of Hypothetical Accident Conditions. However, because the estimated equilibrium temperature of the TPBARs, under Hypothetical Accident Conditions, is expected to be at least 400° F, the >200° F scenario is both bounding, and more realistic, and the ambient to <200° F scenario need not be considered any further.

The design agency's estimate of <55 Ci/TPBAR, in the second column under the heading of *Event-Failed TPBARs*, leads to a total estimated loss of up to 16,500 Ci, or ~0.28 moles of tritium, going directly into containment, for consolidated shipments of up to 300 TPBARs.

In order to calculate the releasable source term for tritium under Hypothetical Accident Conditions, therefore, three different tritium components would have to be considered: 1) the total amount of tritium that had previously been determined above, under Normal Conditions of Transport, in Section 4.5.3.2.1, for *Intact TPBARs*, 2) the total amount of tritium that had previously been determined above, again, under Normal Conditions of Transport, in Section 4.5.3.2.2, for the *Pre-Failed In-Reactor* release scenario, and 3) the total amount of tritium that has just been determined above for Hypothetical Accident Conditions. Should an applicant choose to show all calculations, the reviewer should verify that the releasable source term for tritium, under Hypothetical Accident Conditions, includes all three components.

4.5.5   Leakage Rate Tests for Type B Packages

It was assumed from the outset that the packaging used for the shipment of irradiated TPBARs will be an existing, modified, or newly designed spent-fuel shipping package. Along these lines, it can also be assumed that there will be no fundamental differences between the requirements, and the methodology, used for the fabrication leakage tests for spent-fuel packagings. The same cannot be said for packagings used for the shipment of irradiated TPBARs with respect to the maintenance, and periodic leakage tests, because once a package has been used for the shipment of irradiated TPBARs, the internal surfaces of the package will have been contaminated with tritium. Thus, the procedures used for the maintenance, and periodic leakage tests will have to be conducted in a very different light, because once the internal surfaces of the package have been contaminated with tritium, it can only be assumed that the internal surfaces will always be contaminated with tritium, for the lifetime of the package. Additional precautions will, therefore, have to be built into the procedures used for the maintenance, and periodic leakage tests. (See the additional discussion in Sections 7.5.3 and 8.5.2, below.)

A similar situation also pertains to the requirements for the pre-shipment leakage test. In this case, however, the situation is entirely different, because the pre-shipment leakage test can be designed to comply with the *leaktight* criterion specified in ANSI N14.5, using the closed-loop measurement technique described above. For post-HAC situations, should they become necessary, the closed-loop measurement technique described in Section 7.5.1.2.3 also becomes more important, as this is the only way to determine the amount of tritium "at risk," prior to opening the containment vessel.

4.7   References

4-1.   U.S. Nuclear Regulatory Commission, Spent Fuel Project Office, *Materials Evaluation*, Interim Staff Guidance-15, January 10, 2001.

[†††]   By definition, the term, *Event-Failed TPBARs*, is intended to address the performance of the TPBARs subjected to the conditions during, and after, the Hypothetical Accident.

4-2.   Monroe, R.E., Woo, H.H., and Sears, R.G., Lawrence Livermore National Laboratory, *Recommended Welding Criteria for Use in the Fabrication of Shipping Containers for Radioactive Materials*, NUREG/CR-3019, U.S. Nuclear Regulatory Commission, 1984.

4-3.   Institute for Nuclear Materials Management, *American National Standard for Radioactive Materials—Leakage Tests on Packages for Shipment*, ANSI N14.5-1997, New York, NY, 1998.

4-4.   U.S. Nuclear Regulatory Commission, *Containment Analysis for Type B Packages Used to Transport Various Contents*, NUREG/CR-6487, U.S. Government Printing Office, Washington, D.C., 1996.

4-5.   Knoll, R.W. and Gilbert, E.R., *Evaluation of Cover Gas Impurities and Their Effects on the Dry Storage of LWR Spent Fuel*, PNL-6365, Pacific Northwest National Laboratory, Richland, Washington, November 1987.

4-6.   Pacific Northwest National Laboratory, Tritium Technology Program, *Unclassified Bounding Source Term, Radionuclide Concentrations, Decay Heat, and Dose Rates for the Production TPBAR*, TTQP-1-111, Revision 4, September 16, 2004.

4-7.   Pacific Northwest National Laboratory, Tritium Technology Program, *Unclassified TPBAR Releases, Including Tritium*, TTQP-1-091, Revision 9, May 19, 2004.

4-8.   Westinghouse Electric Company, LLC, *Implementation and Utilization of Tritium-Producing Burnable Absorber Rods (TPBARs) in Watts Bar Unit 1*, NDP-00-0344, Revision 1, July 2001. (See, in particular, Section 3.5, *TPBAR Performance*.)

5.0   SHIELDING REVIEW

5.5   Review Procedures

The shielding review section of NUREG-1609 would normally be applicable to the review of any packaging used for the shipment of irradiated Tritium-Producing Burnable Absorber Rods (TPBARs). However, because TPBARs function in the reactor core like any other burnable poison rods, the shipment of irradiated TPBARs can be expected to take on all of the shielding considerations of shipping containers for spent nuclear fuel; therefore, the shielding review section of NUREG-1617 is also applicable.[5-1]

It is assumed that the packaging to be used will be an existing, modified, or newly designed spent-fuel shipping package. For purposes of this report, however, no specific packaging has been identified for the shipment of such contents. This report, therefore, should be considered to be a topical report, as opposed to a package-specific report.

This section considers each of the subsections of Section 5.5 (Review Procedures) of NUREG-1609 and highlights the special considerations or attention needed for TPBAR shipping packages. In subsections where no significant differences were found, that particular subsection has been omitted from this section. A similar situation also pertains to Section X.5.2.6 of the Spent Fuel Project Office's Interim Staff Guidance document, ISG-15,[5-2] i.e., where no significant differences were found, that particular subsection has been omitted from this section.

For all packages, the shielding review is based in part on the descriptions and evaluations presented in the General Information, Structural Evaluation, Thermal Evaluation, Containment Evaluation, Criticality Evaluation, Operating Procedures, and Acceptance Tests and Maintenance Program sections of the SAR. Similarly, the results of the shielding review are considered in the review of the SAR sections on General Information, Structural Evaluation, Thermal Evaluation, Containment Evaluation, Criticality Evaluation, Operating Procedures, and Acceptance Tests and Maintenance Program.

5.5.2   Radiation Source

5.5.2.1 Gamma Source

In general, the review of the gamma source for irradiated TPBARs should follow the guidance provided in NUREG-1617 for spent nuclear fuel. The key difference between irradiated TPBARs and spent nuclear fuel is that TPBARs have no fissile material; consequently, the gamma source will consist entirely of photons from activated hardware. Because tritium is a low-energy beta emitter, tritium will not contribute to the gamma source.[‡‡‡]

The gamma source term may be calculated using the computer codes ORIGEN-S,[5-3] ORIGEN2,[5-4] or other similar codes. As with any calculations using such codes, the reviewer should follow the guidance in NUREG-1617, Section 5.5.2.1, to verify that the input parameters are applicable to the contents in the application. As stated in NUREG-1617, the input parameters to be reviewed include 1) the ranges of fuel type, burnup, enrichment, and cooling time; 2) the initial composition and mass of the hardware, including impurities, such as ^{59}Co, which form the activation products that are the major contributors to the dose rate; and 3) the spatial and energy variation of the neutron flux during irradiation.

The design agency for the TPBARs (Pacific Northwest National Laboratory [PNNL]) performed unclassified bounding estimates of radionuclide concentrations and the photon source term for irradiated production TPBARs.

[‡‡‡] For purposes of completeness, it should be noted that a continuous spectrum of bremsstrahlung radiation, up to the maximum tritium beta energy of 18.6 keV, will be produced as the beta particles are slowed down in the TPBARs. However, for the spent-fuel package(s) used for the shipment of TPBARs, only photons exceeding approximately 800 keV will contribute significantly to the external radiation levels, so the bremsstrahlung radiation from tritium beta particles may be neglected.

Those estimates are reproduced below in Table 5-1[5-4] and Table 5-2.[5-5] According to Reference 5-5, these results bound the irradiation of production TPBARs in any anticipated host reactor. The calculations considered all components of the TPBARs, and bound all TPBAR designs, including the full-length getter design. Note that the tritium concentrations in Table 5-1 are not the results calculated by ORIGEN2, but rather correspond to the functional requirement of 1.2 grams of tritium (maximum), per TPBAR, corrected for the specified decay times.

**Table 5-1. Maximum Radionuclide Concentrations
in a TPBAR (Ci/TPBAR)**

Nuclide	7 Days	30 Days	90 Days	180 Days	1 Year	5 Years	10 Years
^{3}H	1.16E+04	1.15E+04	1.14E+04	1.13E+04	1.10E+04	8.76E+03	6.61E+03
^{14}C	1.42E-03	1.42E-03	1.42E-03	1.42E-03	1.42E-03	1.42E-03	1.42E-03
^{24}Na	1.98E-02	1.65E-13	0.00E+00	0.00E+00	0.00E+00	0.00E+00	0.00E+00
^{32}P	1.03E+00	3.38E-01	1.84E-02	2.35E-04	3.02E-08	5.78E-10	5.75E-10
^{35}S	1.37E-02	1.15E-02	7.15E-03	3.52E-03	8.18E-04	8.22E-09	4.65E-15
^{37}Ar	3.79E-01	2.40E-01	7.32E-02	1.23E-02	3.15E-04	8.74E-17	1.76E-32
^{39}Ar	9.49E-03	9.49E-03	9.48E-03	9.48E-03	9.46E-03	9.37E-03	9.25E-03
^{42}K	2.18E-04	8.34E-12	8.31E-12	8.27E-12	8.18E-12	7.52E-12	6.77E-12
^{41}Ca	7.51E-05	7.51E-05	7.51E-05	7.51E-05	7.51E-05	7.51E-05	7.51E-05
^{45}Ca	3.13E-01	2.84E-01	2.20E-01	1.50E-01	6.82E-02	1.37E-04	5.78E-08
^{47}Ca	1.57E-04	4.66E-06	4.86E-10	5.17E-16	2.62E-28	0.00E+00	0.00E+00
^{46}Sc	8.20E-03	6.78E-03	4.13E-03	1.96E-03	4.24E-04	2.39E-09	6.57E-16
^{47}Sc	5.68E-04	1.76E-05	1.86E-09	1.98E-15	1.00E-27	0.00E+00	0.00E+00
^{51}Cr	9.67E+02	5.44E+02	1.21E+02	1.28E+01	1.24E-01	1.66E-17	2.38E-37
^{54}Mn	4.19E+01	3.98E+01	3.48E+01	2.85E+01	1.89E+01	7.41E-01	1.29E-02
^{55}Fe	2.15E+02	2.12E+02	2.03E+02	1.90E+02	1.66E+02	5.71E+01	1.51E+01
^{59}Fe	1.98E+01	1.39E+01	5.52E+00	1.38E+00	7.96E-02	1.34E-11	8.14E-24
^{58}Co	2.69E+02	2.15E+02	1.19E+02	4.95E+01	8.06E+00	4.92E-06	8.41E-14
^{60}Co	3.60E+01	3.57E+01	3.49E+01	3.38E+01	3.16E+01	1.87E+01	9.68E+00
^{59}Ni	1.68E-01	1.68E-01	1.68E-01	1.68E-01	1.68E-01	1.68E-01	1.68E-01
^{63}Ni	2.29E+01	2.29E+01	2.28E+01	2.28E+01	2.27E+01	2.20E+01	2.12E+01
^{66}Ni	1.52E-04	1.38E-07	1.59E-15	1.97E-27	0.00E+00	0.00E+00	0.00E+00
^{64}Cu	1.27E-03	1.04E-16	0.00E+00	0.00E+00	0.00E+00	0.00E+00	0.00E+00
^{66}Cu	1.52E-04	1.38E-07	1.59E-15	1.97E-27	0.00E+00	0.00E+00	0.00E+00
^{65}Zn	4.13E-03	3.87E-03	3.26E-03	2.52E-03	1.49E-03	2.34E-05	1.31E-07
^{76}As	8.74E-01	4.25E-07	1.44E-23	0.00E+00	0.00E+00	0.00E+00	0.00E+00
^{75}Se	8.88E-01	7.77E-01	5.49E-01	3.26E-01	1.12E-01	2.38E-05	6.13E-10
^{82}Br	1.14E-03	2.25E-08	1.18E-20	0.00E+00	0.00E+00	0.00E+00	0.00E+00
^{89}Sr	7.51E-02	5.48E-02	2.40E-02	6.99E-03	5.49E-04	1.07E-12	1.39E-23
^{89m}Y	5.48E-04	4.18E-06	1.24E-11	6.39E-20	0.00E+00	0.00E+00	0.00E+00
^{90}Y	5.14E-01	1.30E-03	1.38E-06	1.37E-06	1.36E-06	1.23E-06	1.09E-06
^{91}Y	1.92E-01	1.46E-01	7.19E-02	2.47E-02	2.76E-03	8.38E-11	3.36E-20
^{89}Zr	5.49E-04	4.18E-06	1.25E-11	6.40E-20	5.60E-37	0.00E+00	0.00E+00
^{93}Zr	1.13E-04	1.13E-04	1.13E-04	1.13E-04	1.13E-04	1.13E-04	1.13E-04
^{95}Zr	6.57E+01	5.12E+01	2.67E+01	1.01E+01	1.36E+00	1.81E-07	4.63E-16
^{97}Zr	1.12E-01	1.65E-11	0.00E+00	0.00E+00	0.00E+00	0.00E+00	0.00E+00
^{92}Nb	3.04E-01	6.34E-02	1.06E-03	2.28E-06	7.41E-12	0.00E+00	0.00E+00
^{93m}Nb	3.68E-06	4.02E-06	4.87E-06	6.15E-06	8.73E-06	2.69E-05	4.49E-05
^{94}Nb	4.76E-04	4.76E-04	4.76E-04	4.76E-04	4.76E-04	4.76E-04	4.76E-04
^{95}Nb	6.93E+01	6.50E+01	4.45E+01	1.99E+01	2.94E+00	4.02E-07	1.03E-15

**Table 5-1. Maximum Radionuclide Concentrations
in a TPBAR (Ci/TPBAR)
(Contd.)**

28

Nuclide	7 Days	30 Days	90 Days	180 Days	1 Year	5 Years	10 Years
^{95m}Nb	4.80E-01	3.80E-01	1.98E-01	7.48E-02	1.01E-02	1.34E-09	3.44E-18
^{96}Nb	1.20E-03	9.19E-11	2.51E-29	0.00E+00	0.00E+00	0.00E+00	0.00E+00
^{97}Nb	1.13E-01	1.78E-11	0.00E+00	0.00E+00	0.00E+00	0.00E+00	0.00E+00
^{97m}Nb	1.06E-01	1.57E-11	0.00E+00	0.00E+00	0.00E+00	0.00E+00	0.00E+00
^{93}Mo	1.04E-03	1.04E-03	1.04E-03	1.04E-03	1.04E-03	1.04E-03	1.04E-03
^{99}Mo	1.68E+01	5.11E-02	1.38E-08	1.94E-18	0.00E+00	0.00E+00	0.00E+00
^{99}Tc	4.35E-05	4.36E-05	4.36E-05	4.36E-05	4.36E-05	4.36E-05	4.36E-05
^{103}Ru	3.21E-03	2.14E-03	7.41E-04	1.52E-04	5.76E-06	3.67E-17	3.71E-31
^{115}Cd	2.91E-04	2.27E-07	1.78E-15	1.23E-27	0.00E+00	0.00E+00	0.00E+00
^{115m}Cd	1.84E-04	1.28E-04	5.05E-05	1.25E-05	7.00E-07	9.62E-17	4.52E-29
^{113m}In	1.31E+00	1.14E+00	7.94E-01	4.62E-01	1.51E-01	2.28E-05	3.83E-10
^{114}In	1.26E-01	9.13E-02	3.94E-02	1.12E-02	8.36E-04	1.10E-12	8.64E-24
^{114m}In	1.32E-01	9.54E-02	4.12E-02	1.17E-02	8.73E-04	1.15E-12	9.03E-24
^{113}Sn	1.31E+00	1.14E+00	7.93E-01	4.61E-01	1.51E-01	2.28E-05	3.82E-10
^{117m}Sn	8.21E+00	2.63E+00	1.35E-01	1.57E-03	1.64E-07	0.00E+00	0.00E+00
^{119m}Sn	8.42E+00	7.89E+00	6.66E+00	5.16E+00	3.06E+00	4.90E-02	2.80E-04
^{121}Sn	7.39E-02	4.66E-08	3.12E-24	0.00E+00	0.00E+00	0.00E+00	0.00E+00
^{121m}Sn	5.54E-04	5.53E-04	5.52E-04	5.50E-04	5.46E-04	5.17E-04	4.82E-04
^{123}Sn	4.78E-01	4.22E-01	3.06E-01	1.89E-01	6.99E-02	2.75E-05	1.52E-09
^{125}Sn	2.20E+00	4.21E-01	5.63E-03	8.71E-06	1.43E-11	0.00E+00	0.00E+00
^{122}Sb	1.10E-01	2.99E-04	6.12E-11	5.66E-21	0.00E+00	0.00E+00	0.00E+00
^{124}Sb	1.86E-02	1.43E-02	7.16E-03	2.54E-03	3.01E-04	1.49E-11	1.10E-20
^{125}Sb	1.67E+00	1.66E+00	1.60E+00	1.50E+00	1.32E+00	4.87E-01	1.39E-01
^{126}Sb	5.64E-02	1.56E-02	5.45E-04	3.55E-06	1.13E-10	0.00E+00	0.00E+00
^{123m}Te	3.02E-03	2.65E-03	1.87E-03	1.11E-03	3.80E-04	8.02E-08	2.05E-12
^{125m}Te	3.26E-01	3.40E-01	3.58E-01	3.56E-01	3.22E-01	1.19E-01	3.40E-02
^{131}Cs	5.10E-02	2.34E-02	1.17E-03	7.33E-06	1.50E-10	0.00E+00	0.00E+00
^{131}Ba	3.68E-02	9.53E-03	2.81E-04	1.43E-06	2.69E-11	0.00E+00	0.00E+00
^{133}Ba	7.43E-04	7.40E-04	7.32E-04	7.20E-04	697E-04	5.38E-04	3.90E-04
^{133m}Ba	3.65E-05	1.95E-09	1.39E-20	2.26E-37	0.00E+00	0.00E+00	0.00E+00
^{135m}Ba	2.77E-04	4.49E-10	3.51E-25	0.00E+00	0.00E+00	0.00E+00	0.00E+00
^{140}La	3.92E-04	1.86E-07	6.07E-09	4.62E-11	2.02E-15	0.00E+00	0.00E+00
^{177}Lu	2.13E-03	1.99E-04	1.57E-06	7.79E-07	3.40E-07	4.95E-10	1.40E-13
^{175}Hf	3.25E-02	2.59E-02	1.43E-02	5.86E-03	9.37E-04	4.88E-10	6.84E-18
^{181}Hf	8.82E-01	6.06E-01	2.27E-01	5.22E-02	2.52E-03	1.07E-13	1.15E-26
^{182}Ta	1.07E+01	9.33E+00	6.50E+00	3.78E+00	1.24E+00	1.85E-04	3.84E-09
^{183}Ta	2.54E+01	1.12E+00	3.21E-04	1.56E-09	1.82E-20	0.00E+00	0.00E+00
^{181}W	5.88E-03	5.16E-03	3.66E-03	2.19E-03	7.58E-04	1.78E-07	5.17E-12
^{185}W	2.09E-01	1.69E-01	9.69E-02	4.22E-02	7.64E-03	1.06E-08	5.09E-16
^{187}W	2.68E-02	2.99E-09	2.18E-27	0.00E+00	0.00E+00	0.00E+00	0.00E+00
^{188}W	1.65E-02	1.31E-02	7.22E-03	2.94E-03	4.62E-04	2.12E-10	2.54E-18
^{186}Re	3.18E-02	4.66E-04	7.70E-09	5.16E-16	8.85E-31	0.00E+00	0.00E+00
^{188}Re	1.79E-02	1.33E-02	7.29E-03	2.97E-03	4.67E-04	2.15E-10	2.57E-18
^{191}Os	4.87E-05	1.73E-05	1.16E-06	2.03E-08	4.86E-12	0.00E+00	0.00E+00
Nuclide	7 Days	30 Days	90 Days	180 Days	1 Year	5 Years	10 Years
Totals	1.34E+04	1.28E+04	1.21E+04	1.17E+04	1.12E+04	8.86E+03	6.66E+03

Note: Adapted from Reference 5-4.

Table 5-2. Maximum Photon Source Term
in a TPBAR (Photons/(TPBAR·s))

Energy (MeV)	7 Days	30 Days	90 Days	180 Days	1 Year	5 Years	10 Years
1.00E-02	7.73E+12	5.07E+12	2.33E+12	1.14E+12	6.01E+11	3.17E+11	2.28E+11
2.50E-02	6.71E+11	4.15E+11	2.59E+11	1.76E+11	1.03E+11	1.95E+10	7.02E+09
3.75E-02	1.80E+11	1.08E+11	6.65E+10	3.72E+10	1.85E+10	6.83E+09	2.84E+09
5.75E-02	5.80E+11	4.44E+11	2.90E+11	1.60E+11	5.27E+10	4.20E+09	2.15E+09
8.50E-02	1.52E+11	9.81E+10	5.86E+10	2.93E+10	9.11E+09	1.66E+09	8.49E+08
1.25E-01	2.24E+11	1.41E+11	8.80E+10	4.66E+10	1.45E+10	7.08E+08	3.45E+08
2.25E-01	4.52E+11	2.38E+11	1.20E+11	6.46E+10	2.15E+10	1.30E+09	4.20E+08
3.75E-01	3.06E+12	1.73E+12	4.10E+11	6.55E+10	1.94E+10	6.57E+09	1.90E+09
5.75E-01	2.75E+12	2.17E+12	1.21E+12	5.16E+11	1.02E+11	8.36E+09	2.39E+09
8.50E-01	1.56E+13	1.29E+13	7.83E+12	3.77E+12	1.11E+12	2.70E+10	5.28E+08
1.25E+00	3.05E+12	2.96E+12	2.81E+12	2.63E+12	2.38E+12	1.38E+12	7.16E+11
1.75E+00	5.01E+10	3.96E+10	2.20E+10	9.10E+09	1.48E+09	9.09E+02	5.52E+00
2.25E+00	2.12E+09	3.75E+08	3.27E+07	1.84E+07	1.30E+07	7.33E+06	3.80E+06
2.75E+00	7.48E+08	6.48E+04	5.30E+04	4.48E+04	3.88E+04	2.27E+04	1.18E+04
3.50E+00	5.05E+05	1.88E+00	6.13E-02	4.70E-04	3.16E-06	2.87E-06	2.58E-06
5.00E+00	5.21E+03	5.25E-08	6.64E-09	4.23E-09	1.67E-09	1.11E-12	1.93E-15
7.00E+00	6.37E-10	5.81E-10	4.31E-10	2.75E-10	1.09E-10	7.23E-14	1.25E-16
9.50E+00	4.03E-11	3.68E-11	2.72E-11	1.74E-11	6.87E-12	4.57E-15	7.93E-18
Energy (MeV)	7 Days	30 Days	90 Days	180 Days	1 Year	5 Years	10 Years
Totals	3.45E+13	2.63E+13	1.55E+13	8.65E+12	4.44E+12	1.78E+12	9.63E+11

Note: Adapted from Reference 5-5.

The photon source terms shown in Table 5-2 above are given as functions of energy group and decay time (i.e., time since the end of irradiation). Earlier decay times correspond to larger photon source terms; therefore, the photon source term will be conservative if the decay time of the photon source term used in the shielding evaluation is less than the decay time of the TPBARs to be shipped. Because the decay time assumed in the shielding evaluation becomes a condition of approval in the certificate of compliance, the applicant should ensure that it accommodates their shipping requirements.

According to the information presented in Reference 5-6, a decay time of 30 days should be sufficiently conservative for the photon source term in the shielding evaluation, based on the following:

> "About 30 days after the refueling is complete, plant operators would begin to remove the remaining irradiated TPBAR assemblies from the spent fuel assemblies, disassemble all of the irradiated TPBARs for consolidation, and place them into consolidation canisters. The time to start consolidating the TPBARs is not limited by any safety issues (e.g., decay heat), but rather is based on scheduling. The 30-day estimate corresponds to when the licensee expects to be finished with all outage-related activities, and can begin consolidation efforts."[5-6]

5.5.2.2 Neutron Source

This section is not applicable for the shipment of irradiated TPBARs, as the TPBARs do not produce neutrons.

5.5.4 Shielding Evaluation

Other than not including a neutron source term in the calculations along with a photon source term, there should be no significant differences in the general methods provided in NUREG-1617 for the review of spent-fuel packages. The one exception is that a minimum cooling time of 30 days should be imposed on the shipment of irradiated TPBARs, as per the information provided in References 5-5 and 5-6.

5.7 References

5-1. U.S. Nuclear Regulatory Commission, *Standard Review Plan for Transportation Packages for Spent Nuclear Fuel*, NUREG-1617, U.S. Government Printing Office, Washington, D.C., 1999.

5-2.   U.S. Nuclear Regulatory Commission, Spent Fuel Project Office, *Materials Evaluation*, Interim Staff Guidance-15, January 10, 2001.

5-3.   Radiation Safety Information Computational Center, *SCALE 5: Modular Code System for Performing Standardized Computer Analyses for Licensing Evaluation for Workstations and Personal Computers*, Code Package CCC-725, Oak Ridge National Laboratory, June 2004.

5-4.   Radiation Safety Information Computational Center, *ORIGEN2 V2.2: Isotope Generation and Depletion Code Matrix Exponential Method*, Code Package CCC-371, Oak Ridge National Laboratory, June 2002.

5-5.   Pacific Northwest National Laboratory, Tritium Technology Program, *Unclassified Bounding Source Term, Radionuclide Concentrations, Decay Heat, and Dose Rates for the Production TPBAR*, TTQP-1-111, Revision 4, September 16, 2004.

5-6.   U.S. Nuclear Regulatory Commission, *Safety Evaluation by the Office of Nuclear Reactor Regulation Related to Amendment No. 40 to Facility Operating License No. NPF-90 Tennessee Valley Authority Watts Bar Nuclear Plant, Unit 1 Docket No. 50-390*, September 23, 2002. (See, in particular, Section 2.1.1.) Note: This particular document was included as Enclosure 2, as part of a letter, L. M. Padovan (NRC), to J. A. Scalice (TVA), dtd., September 23, 2002, Subject: Watts Bar Nuclear Plant, Unit 1—Issuance of Amendment to Irradiate up to 2,304 Tritium-Producing Burnable Absorber Rods in the Reactor Core (TAC NO. MB1884), ADAMS Accession No. ML022540925.

6.0   CRITICALITY REVIEW

6.5   Review Procedures

The criticality review section of NUREG-1609 would normally be applicable to the review of any packaging used for the shipment of irradiated Tritium-Producing Burnable Absorber Rods (TPBARs). For purposes of this report, however, no specific packaging has been identified for the shipment of such contents. This report, therefore, should be considered to be a topical report, as opposed to a package-specific report.

It is assumed that the packaging to be used will be an existing, modified, or newly designed spent-fuel shipping package. However, because the contents of the package will contain no fissile material, the review format will follow that specified in NUREG-1609.

This section considers each of the subsections of Section 6.5 (Review Procedures) of NUREG-1609 and highlights the special considerations or attention needed for TPBAR shipping packages. In subsections where no significant differences were found, that particular subsection has been omitted from this section. A similar situation also pertains to Section X.5.2.7 of the Spent Fuel Project Office's Interim Staff Guidance document, ISG-15,[6-1] i.e., where no significant differences were found, that particular subsection has been omitted from this section.

For all packages, the criticality review is based in part on the descriptions and evaluations presented in the General Information, Structural Evaluation, Thermal Evaluation, Containment Evaluation, Shielding Evaluation, Operating Procedures, and Acceptance Tests and Maintenance Program sections of the SAR. Similarly, the results of the criticality review are considered in the review of the SAR sections on General Information, Structural Evaluation, Thermal Evaluation, Containment Evaluation, Shielding Evaluation, Operating Procedures, and Acceptance Tests and Maintenance Program.

6.5.2   Fissile Material Contents

No fissile material contents are associated with the shipment of irradiated TPBARs. There are, therefore, no criticality concerns.

6.7   References

6-1.   U.S. Nuclear Regulatory Commission, Spent Fuel Project Office, *Materials Evaluation*, Interim Staff Guidance-15, January 10, 2001.

7.0 OPERATING PROCEDURES REVIEW

7.5 Review Procedures

The operating procedures review section of NUREG-1609 would normally be applicable to the review of any packaging used for the shipment of irradiated Tritium-Producing Burnable Absorber Rods (TPBARs). For purposes of this report, however, no specific packaging has been identified for the shipment of such contents. This report, therefore, should be considered to be a topical report, as opposed to a package-specific report.

It is assumed that the packaging to be used will be an existing, modified, or newly designed spent-fuel shipping package. However, because the contents of the package will contain no fissile material, the review format will follow that specified in NUREG-1609.

This section considers each of the subsections of Section 7.5 (Review Procedures) of NUREG-1609 and highlights the special considerations or attention needed for TPBAR shipping packages. In subsections where no significant differences were found, that particular subsection has been omitted from this section.

For all packages, the operating procedures review is based in part on the descriptions and evaluations presented in the General Information, Structural Evaluation, Thermal Evaluation, Containment Evaluation, Shielding Evaluation, Criticality Evaluation, and Acceptance Tests and Maintenance Program sections of the SAR. Similarly, the results of the operating procedures review are considered in the review of the SAR sections on General Information, Structural Evaluation, Thermal Evaluation, Containment Evaluation, Shielding Evaluation, Criticality Evaluation, and Acceptance Tests and Maintenance Program.

7.5.1 Package Loading

The reviewer should verify that, prior to the start of any work with irradiated TPBARs, provisions are in place for the real-time monitoring of tritium in air. The reviewer should also verify that additional provisions are in place for the sampling of tritium in water, particularly the water in the spent-fuel pool, and the water in the cask during the vacuum-drying process. The reviewer should then verify that provisions are in place for the follow-up sampling of tritium contamination levels in the vacuum pump oils that will become contaminated as part of the vacuum-drying processes used after loading. Finally, the reviewer should verify that provisions are in place for the measurement of basic tritium surface-contamination levels. (Note that most of these provisions will be very different from those normally encountered in typical reactor operations environments. See Appendix F.)

Also, because there is the very real possibility that workers could be exposed to tritium levels that are not normally associated with reactor work, the reviewer should verify that the operating procedures clearly state that all personnel involved with TPBAR loading operations will be on a tritium bioassay program, in accordance with Regulatory Guide 8.32.[7-1]

7.5.1.1 Preparation for Loading

The reviewer should verify that the special controls and precautions noted above are included, i.e., having appropriate tritium monitoring capabilities in place prior to beginning preparation for loading. The reviewer should also verify that additional procedures are in place to deal specifically with the determination of residual tritium outgassing and contamination in any package that has previously been used for TPBAR transport, and appropriate precautions are in place to notify the user that tritium releases are possible when opening an "empty" cask and, possibly, during other cask operations.

The reviewer should further verify that no elastomeric seals are used in any part of the containment boundary.[§§§]

[§§§] For purposes of this document, the term *elastomeric seals* pertains equally to organic, elastomeric, halogenated hydrocarbon, thermoplastic resin, and/or thermosetting resin types of seals. See Section 4.5.1.1.3; see also Appendix E.

7.5.1.2 Loading of Contents

As was noted above in Sections 3.5.4 and 4.5.2.3, the transport package for irradiated TPBARs will be loaded under water. It was also noted that the package will be vacuum dried and backfilled with an inert gas, in accordance with the generic procedures outlined in the Pacific Northwest National Laboratory document, *Evaluation of Cover Gas Impurities and Their Effects on the Dry Storage of LWR Spent Fuel*.[7-2] But, because Reference 7-2 does not address tritium-specific issues, the reviewer should verify that the appropriate tritium health physics considerations outlined in Sections 7.5.1.2.1, 7.5.1.2.2, and 7.5.1.2.3, below, are included.

7.5.1.2.1   Contaminated Water Issues

It should be assumed from the outset that the water from the spent-fuel pool and the cask-loading pit will be contaminated with tritium, possibly up to several tens of $\mu Ci/ml$.[7-3] As such, there should be a cautionary note in the procedures stating, in effect, that contact with water from the spent-fuel pool and/or the cask-loading pit should be avoided to the maximum extent possible. Should a worker be splashed with water from either the spent-fuel pool or the cask-loading pit, the contaminated water should be washed off with clean water immediately. This will help minimize the potential dose to the worker. (See Appendix F.)

It should also be noted that, because the water in the cask will have come from the spent-fuel pool/cask-loading pit, the water in the cask will also be tritium contaminated. But, it should not necessarily be expected that the contamination levels in the cask water will be the same as that in the spent-fuel pool/cask-loading pit. The tritium contamination levels in the cask will be dependent on the physical condition of the TPBARs (i.e., intact TPBARs vs. event-failed TPBARs) and the total permeation loss rate from the consolidated batch.[****] Since the volume of the water in the cask is much smaller than the volume of water in the spent-fuel pool/cask-loading pit, the tritium contamination levels in the cask water could easily be higher—substantially higher—than the tritium contamination levels in the spent-fuel pool/cask-loading pit. As a consequence, therefore, the same precautions that applied above with respect to splashing with water from the spent-fuel pool/cask-loading pit apply equally to the case of splashing with drainage water from the cask, i.e., should a worker be splashed with cask-drainage water, the contaminated water should be washed off with clean water immediately.

In order to better understand the potential hazards from splashing with water from the spent-fuel pool, the cask-loading pit, and/or the cask-drainage water, it is recommended that samples be taken, early and often, throughout the cask-draining process. Such samples can be analyzed, through the use of the liquid-scintillation counting, to determine the relative hazard potential, at any point in time.

7.5.1.2.2   Contaminated Vapor Issues

Once the bulk of the water has been removed from the cask interior, the process of vacuum drying can begin. Here, too, additional precautions must be taken, because the exhaust gases and vapors from the vacuum-drying equipment will be tritium contaminated. As an immediate consequence, the procedures used must include provisions for the proper venting of the exhaust gases, so that they will not be vented directly into the room or into the breathing zone of the workers. As a follow-up consequence, it should also be noted that the pump oils used in the vacuum-drying system will also become contaminated with tritium, quite possibly up to several Ci/l. Since direct contact with the pump oil from the vacuum-drying system can represent an additional health physics hazard, contact with the vacuum pump oils and vapors should also be avoided.

Because the equipment used in the vacuum-drying process for irradiated TPBARs has the potential to be tritium contaminated, and because the tritium levels in some parts of the equipment can be expected to be relatively high, the equipment used for the vacuum-drying process for irradiated TPBARs should *not* be used for the vacuum drying of any other packages. Potential options should include decontamination of the equipment internals, changing of the vacuum pump oils to levels that indicate that the pump oils are no longer tritium contaminated, and/or dedicated storage of such equipment for use only for shipments of irradiated TPBARs.

[****] See also the discussion above, on permeation loss rates, in Section 4.5.3.

7.5.1.2.3   Pre-Shipment TPBAR Outgassing Measurements

Once the internals of the cask have been drained and dried and the cask has been backfilled with the inert gas of choice, one additional set of measurements should be made, one that will help determine the amount of tritium that might be "at risk," at any point in time, during transport. Given the variety of possibilities described above in Table 4-1 and in Section 4.5.3, the measurements described below can be considered to be optional (if the applicant has elected to default to the *leaktight* criteria specified in ANSI N14.5) or mandatory (if the applicant has shown by calculation that some lesser criteria can be used). In either case, however, the measurement techniques described below are the only way to determine the amount of tritium that might be "at risk," at any point in time, during transport.[††††] (Note: If the applicant has shown by calculation that the containment criteria to be used are less than *leaktight*, this is also the only way to verify that the containment criteria defined in Section 4 will not be exceeded for Normal Conditions of Transport.)

Standard practices in the tritium business suggest that no closed containers shall be opened without a preliminary determination of the airborne tritium levels that might be "at risk," i.e., the amount of tritium that might be available to go into, or through, the worker's breathing zone(s), and/or the amount of tritium that might be available to be released directly to the environment. These types of measurements are typically performed with a closed-loop monitoring system that circulates air (or a pre-selected monitoring gas, such as dry nitrogen, helium or argon) into, and out of, the enclosure in question, through a tritium monitor that has the capability of determining "real-time" tritium concentrations. Once the tritium concentration inside the containment vessel has been determined, the total amount of tritium "at risk," at any given time, can be determined.

Once the amount of tritium "at risk" has been determined, at the shipping facility, prior to shipment, the receiving facility can be notified as to what they might expect upon receipt. Once the amount of tritium "at risk" has been determined at the receiving facility, the receiving facility will be able to compare its measurements to those performed previously at the shipping facility. Armed with this kind of information, the receiving facility should have several options in place to deal with the situation, one of which should include the option of running the containment gases through a local clean-up system prior to opening the containment vessel. A second option that should also be considered is the sampling of the containment gases for the actual gas composition, and the subsequent determination of potential combustible-gas mixtures that might be encountered as part of the unloading process.

7.5.1.3 Preparation for Transport

For the most part, the procedures used for this portion of the operating procedures should be similar to those used for the shipment of any other radioactive material, including spent fuel. There are, however, a number of areas where the procedures used could be/should be quite different. Each is described below.

7.5.1.3.1   Pre-Shipment Radiation Surveys

For the shipment of irradiated TPBARs, the pre-shipment dose rate measurement requirements should be virtually identical to the requirements for the shipment of spent nuclear fuel. As was noted in Section 5.5.2.2, however, there should be no production of neutrons from irradiated TPBARs. The pre-shipment requirement for neutron dose rate measurements can, therefore, be eliminated for the shipment of irradiated TPBARs.

7.5.1.3.2   Pre-Shipment Surface Contamination Measurements

For the shipment of irradiated TPBARs, the pre-shipment surface contamination measurement requirements will have to be broken down into two distinct types: 1) routine surface contamination measurements for gross beta-gamma contamination, and 2) routine surface contamination measurements for tritium "outgassing." (See Section E.6.3. See also Appendix E, in general.) Although the former type of measurement is routinely required for the shipment of most radioactive materials, including spent fuel, the phenomenon known as "outgassing" in the tritium business is equivalent to "cask-weeping" in the spent-fuel business.

[††††] See the additional discussion in Sections E.4, E.5, and E.6.

7.5.1.3.3   Pre-Shipment Leakage Tests

For the shipment of most radioactive materials, the requirements of ANSI N14.5 specify that the pre-shipment leakage test will be capable of detecting a volumetric leakage rate to a level of 10^{-3} reference·cm^3/s.[7-4] It is not uncommon, however, in the tritium business, to adopt a pre-shipment leakage test requirement of *leaktight*, as defined in ANSI N14.5. (See Section 4.5.3.) Should an applicant choose to adopt the ANSI N14.5 *leaktight* requirement for the pre-shipment leakage test, it should be verified that the method(s) selected by the applicant can be used to meet the 10^{-7} reference·cm^3 requirement for the pre-shipment leakage test.

7.5.1.3.4   Special Instructions

Under the broader heading of special instructions that should be provided to the consignee for opening the package, the following should be provided as part of the pre-shipment information:

1) The pre-shipment results from the surface contamination measurements for gross beta-gamma contamination;

2) The pre-shipment results from the surface contamination measurements for tritium; and

3) The tritium outgassing levels from the procedures described above in Section 7.5.1.2.3.

7.5.2   Package Unloading

As was noted previously in Section 7.5.1, the reviewer should verify that monitoring and sampling provisions are in place for tritium in any of the forms that might be encountered, i.e., tritium in air, tritium in water, tritium in vacuum pump oils, etc. Because the receiving facility will be the Tritium Extraction Facility (TEF), located at the Department of Energy's (DOE's) Savannah River Site (SRS), it is expected that the tritium monitoring requirements described above will be in place, as specified. Also, because the TEF can be expected to operate along the same lines as any other DOE tritium facility, it is also expected that the personnel involved with the unloading operations will already be on a tritium bioassay program.

7.5.2.1 Receipt of Package from Carrier

The reviewer should verify that the standard radiation survey measurements are taken upon arrival of the package at the receiving facility. As was noted previously, the TPBAR contents do not produce neutrons, so there should be no need for neutron measurements as part of the incoming survey.

For the surface contamination measurements, however, the reviewer should verify that *two,* distinctly different types of surface contamination measurements are required on the external surface of the package, the first being for gross, beta-gamma surface contamination, and the second being for surface contamination measurements for tritium.

7.5.2.2 Removal of Contents

The reviewer should verify that, prior to the removal of the contents, there is a step in the procedures to determine the amount of tritium that might be "at risk," *before* the containment vessel is opened. The method should follow the techniques described above in Section 7.5.1.2.3, and, in this case, the user should be *required* to perform such a measurement, prior to the unloading of TPBARs. Given the variety of possibilities described above in Table 4-1, and in Section 4.5.3, this is the only way that the actual amount of tritium "at risk" can be determined in a real-time, on-the-spot, situation.

Once the amount of tritium "at risk" has been determined at the receiving facility, the receiving facility will be able to compare its measurements against those performed previously at the shipping facility. Armed with this kind of information, the receiving facility should have several options in place to deal with the situation, one of which, as was noted above, includes the option of running the containment gases through a local clean-up system, prior to opening the containment vessel. A second option that should also be available is the sampling of the containment gases for the actual gas composition, and the subsequent determination of potential combustible-gas mixtures that might be encountered as part of the unloading process.

7.5.3 Preparation of Empty Package for Transport/Storage

Whether the package is to be returned to the reactor, or whether the cask is to be placed in storage, once it has been used for the transport of TPBARs, the internal surfaces of the containment vessel will have been contaminated with tritium. As a consequence, the cask can no longer be considered as being *empty*, with respect to its tritium content. Therefore, before the *empty* cask is moved to its next destination, the residual containment vessel gases will have to be sampled again, using the same basic measurement techniques described above in Section 7.5.1.2.3. The purpose of the measurement, in this case, however, is to establish a baseline value for the tritium outgassing rate from the internal surfaces of the containment vessel, from a supposedly *empty* cask.

Similar measurements will have to be repeated again, prior to opening the cask, at the next destination. The purpose of the measurements, in this case, however, is to determine the amount of tritium that might be "at risk," at the new receiving destination. If the amount of tritium that might be "at risk" is on the order of a few, to several tens, to several hundreds of curies, a receiving reactor site may have no objections to discharging that amount of tritium directly into its spent-fuel pool. If, on the other hand, the receiving site is a maintenance facility, where the cask would be opened to room air, amounts of tritium on the order of a few, to several tens, to several hundreds of curies "at risk," discharged directly into the room air, and/or the breathing environment, would probably not be acceptable.

From a regulatory standpoint, it should also be noted that once a container has been used for the shipment of irradiated TPBARs, it can probably, never again, be shipped as an *empty* container. While the measurement techniques described above are sensitive enough to demonstrate that the amount of tritium "at risk" is well below an A_2 value for tritium (i.e., 1,080 Ci), the internal surface contamination limits requirements, specified in 49 CFR 173.428(c) and 49 CFR 173.443(a), now become the limiting factors.[‡‡‡‡] (See also the additional discussion in Sections F.5.1.1.1 and F.5.1.1.3, below.)

Finally, it should be noted that, because it should be expected that residual amounts of tritium will always be present on/in the internal surfaces of the containment vessel, additional maintenance requirements will have to be added to look for signs of intergranular-corrosion and/or intergranular-stress-corrosion cracking over time, particularly if the containment vessel is constructed of materials other than Type 304L or Type 316L stainless steels. (See the additional discussion in Sections 2.5.2, 2.5.3, 2.5.6, 2.5.7, and 4.5.1, above, and Section 8.5.2, below.)

7.7 References

7-1. U.S. Nuclear Regulatory Commission, Regulatory Guide 8.32, *Criteria for Establishing a Tritium Bioassay Program*, U.S. Government Printing Office, July 1988.

7-2. Knoll, R.W. and Gilbert, E.R., *Evaluation of Cover Gas Impurities and Their Effects on the Dry Storage of LWR Spent Fuel*, PNL-6365, Pacific Northwest National Laboratory, Richland, Washington, November 1987.

7-3. Westinghouse Electric Company, LLC, *Implementation and Utilization of Tritium-Producing Burnable Absorber Rods (TPBARs) in Watts Bar Unit 1*, NDP-00-0344, Revision 1, July 2001. (See, in particular, Section 1.5.1, pp. 1-14 through 1-19, and Section 3.7.3, pp. 3-22 through 3-27.)

7-4. Institute for Nuclear Materials Management, *American National Standard for Radioactive Materials— Leakage Tests on Packages for Shipment*, ANSI N14.5-1997, New York, NY, 1998.

[‡‡‡‡] See also the additional discussion in Sections 4.5.3, E.6.1, E.6.2, E.6.3, and E.6.4, and in Appendix E, in general.

8.0 ACCEPTANCE TESTS AND MAINTENANCE PROGRAM REVIEW

8.5 Review Procedures

The acceptance tests and maintenance program review section of NUREG-1609 would normally be applicable to the review of any packaging used for the shipment of irradiated Tritium-Producing Burnable Absorber Rods (TPBARs). For purposes of this report, however, no specific packaging has been identified for the shipment of such contents. This report, therefore, should be considered to be a topical report, as opposed to a package-specific report.

It is assumed that the packaging to be used will be an existing, modified, or newly designed spent-fuel shipping package. However, because the contents of the package will contain no fissile material, the review format will follow that specified in NUREG-1609.

8.5.1 Acceptance Tests

For all packages, the acceptance tests review is based in part on the descriptions and evaluations presented in the General Information, Structural Evaluation, Thermal Evaluation, Containment Evaluation, Shielding Evaluation, Criticality Evaluation, and Operating Procedures sections of the SAR. Similarly, results of the acceptance tests review are also considered in the review of the SAR sections on General Information, Structural Evaluation, Thermal Evaluation, Containment Evaluation, Shielding Evaluation, Criticality Evaluation, and Operating Procedures.

Because it has already been assumed that the packaging to be used for the shipment of irradiated TPBARs will be an existing, modified, or newly designed spent-fuel shipping package, there should be no significant differences in the acceptance test requirements for irradiated TPBAR packages, relative to the requirements for new spent-fuel packages, or new radioactive materials packages.

8.5.2 Maintenance Program

For all packages, the maintenance program review is based in part on the descriptions and evaluations presented in the General Information, Structural Evaluation, Thermal Evaluation, Containment Evaluation, Shielding Evaluation, Criticality Evaluation, and Operating Procedures sections of the SAR. Similarly, results of the maintenance program review are also considered in the review of the SAR sections on General Information, Structural Evaluation, Thermal Evaluation, Containment Evaluation, Shielding Evaluation, Criticality Evaluation, and Operating Procedures.

Note: After the package has been used for the shipment of irradiated TPBARs, it should be assumed that the internals of the package are contaminated with tritium. Prior to opening an *empty* package, therefore, the appropriate precautions should be taken to verify that the internal walls of the containment vessel are not outgassing. (See the related discussion in Sections 4.5.3.2.4 and 7.5.3, above. See also Sections E.4, E.5, and E.6, below.) This type of information can be particularly important to note for leakage testing purposes—to determine the amount of tritium (as HTO) that might have to be pumped through a vacuum system—and as information to be used for pre-inspection purposes—so that the workers can be appropriately notified of potential HTO outgassing problems.

8.5.2.2 Component and Material Tests

As was noted in Section 4.5.3.2, above, it should be expected that the internals of the package will become contaminated with tritium any time the package is used for the shipment of irradiated TPBARs. As part of the maintenance program, therefore, special attention should be paid to potential long-term corrosion issues. At a minimum, therefore, it is recommended that an additional requirement be added to the maintenance program to require an annual inspection by a qualified corrosion metallurgist of all accessible containment surfaces, welds, heat-affected zones, and sealing surfaces for evidence of corrosive attack or residue.

It is further recommended that a record be kept of the total amount of tritium that has been released into the containment vessel for each package used. The total amount of tritium for any given shipment can be determined from the outgassing measurements mandated above in Section 7.5.2.2. Such records should be kept for the lifetime of the package.

APPENDICES

Appendix E: Physical and Chemical
Properties of Tritium

(Note: The bulk of the information presented in this Appendix was adapted from Sections 2.10.1 through 2.10.6 of the Department of Energy's, "Design Considerations," published in 1999.[E-1] Although some of the information may appear to be somewhat dated, the basic concepts behind the information have not change since that time.* See also the information presented in Appendix F.)

E.1   Sources of Tritium

Tritium is the lightest of the naturally occurring radioactive nuclides. Tritium is produced in the upper atmosphere as a result of cascade reactions between incoming cosmic rays and elemental nitrogen. In its simplest form, this type of reaction can be written as

$$^{14}_{7}\text{N} + ^{1}_{0}n \rightarrow ^{12}_{6}\text{C} + ^{3}_{1}\text{H}. \tag{E.1}$$

Tritium is also produced in the sun as a sub-set of the proton-proton chain of fusion reactions. Although a steady stream of the tritium near the surface of the sun is ejected out into space (along with many other types of particles) on the solar wind, much larger streams are ejected out into space during solar flares and prominences. Being much more energetic than its solar wind counterpart, tritium produced in this manner is injected directly into the earth's upper atmosphere as the earth moves along in its orbit. Regardless of the method of introduction, however, estimates suggest that the natural production rate for tritium is about 4×10^6 Ci/yr, which, in turn, results in a steady-state, natural production inventory of about 7×10^7 Ci.

Tritium is also introduced into the environment through a number of man-made sources. The largest of these, atmospheric nuclear testing, added approximately 8×10^9 Ci to the environment between 1945 and 1975. Because the half-life of tritium is relatively short, i.e., about 12.3 years—see Section E.3.1, below, much of the tritium produced in this manner has long since decayed. However, tritium introduced into the environment as a result of atmospheric testing increased the natural background levels by more than two orders of magnitude, and, in spite of its relatively short half-life, the natural background levels of tritium in the environment will not return to normal until sometime between the years 2020 and 2030.

Tritium levels in the environment cannot truly return to background levels, however, because of a number of additional man-made sources. Tritium is also produced as a ternary fission product, within the fuel rods of nuclear reactors, at a rate of $1–2 \times 10^4$ Ci/1,000 MW(e). (Although much of the tritium produced in this manner remains trapped within the matrix of the fuel rods, estimates suggest that recovery of this tritium could reach levels of 1×10^6 Ci/yr.) Typical light-water and heavy-water moderated reactors produce another 500–1,000 to 1×10^6 Ci/yr, respectively, for each 1,000 MW of electrical power. Commercial producers of radio-luminescent and neutron generator devices also release about 1×10^6 Ci/yr. Thus, tritium facilities operate within a background of tritium from a variety of sources.

E.2   The Relative Abundance of Tritium

The isotopes of hydrogen have long been recognized as being special—so special, in fact, that each has been given its own chemical name and symbol. Protium, for example, is the name given to the hydrogen isotope of mass-1, and the chemical symbol for protium is H. Deuterium is the name given to the hydrogen isotope of

---

*   Additional Note: Because the bulk of the information presented in this Appendix is presented in a paraphrased format, it is suggested that the reader refer directly to Reference E-1 for additional information, which does include all of the appropriate references to the original citations.

mass-2; the chemical symbol for deuterium is D. Tritium is the name given to the hydrogen isotope of mass-3. Its chemical symbol is T.

Protium is by far the most abundant of the hydrogen isotopes. Deuterium follows next with a relative abundance of about 1 atom of deuterium for every 6,600 atoms of protium; that is, the D to H ratio (D:H) is about 1:6,600. Tritium is the least common hydrogen isotope. The relative abundance of naturally occurring tritium (i.e., tritium produced in the upper atmosphere and tritium injected directly by the sun) has been estimated to be on the order of 1 tritium atom for every 10^{18} protium atoms. The introduction of man-made tritium into the environment, particularly as a result of atmospheric testing, has raised this level approximately one order of magnitude so that the ambient T to H ratio is now approximately $1:10^{17}$.

The names, commonly used chemical and nuclear symbols, atomic masses, and relative natural abundances of the hydrogen isotopes are summarized below in Table E-1.

Table E-1.
The Isotopes of Hydrogen.

Name	Chemical Symbol	Nuclear Symbol	Atomic Mass	Natural Abundance (%)	Natural Abundance (x:H Ratio)
Protium	H	$^{1}_{1}H$	1.007 825 03	99.985 %	1:1
Deuterium	D	$^{2}_{1}H$	2.014 101 78	0.015 %	1:6,600
Tritium	T	$^{3}_{1}H$	3.016 049 26*	Very Low	$1:10^{17}$

* Calculated

E.3   Radioactive Decay of Tritium

E.3.1   Generic

As the lightest of the pure beta emitters, tritium decays with the emission of a low energy beta particle and an anti-neutrino, i.e.,

$$^{3}_{1}H \rightarrow \, ^{3}_{2}He + \beta^{-} + \overline{\upsilon}. \tag{E.2}$$

Tritium decays with a half-life of 12.32 years. The specific activity of tritium is approximately 9,619 Ci/gram, and/or 1.040×10^{-4} g/Ci. In addition, the activity density (i.e., the specific activity per unit volume) for tritium gas (T_2) is 2.589 Ci/cm^3, under standard temperature and pressure (STP) conditions (i.e., 1 atmosphere of pressure at 0° C), and/or 2.372 Ci/cm^3 at 25°C. It can also be shown that the former value translates to 58,023 Ci/gm-mole and 29,012 Ci/gm-atom, under STP conditions.

E.3.2   Beta Emissions

Beta particles interact with matter by colliding with bound electrons in the surrounding medium. In each collision, the beta particle loses energy as electrons are stripped from molecular fragments (ionization) or promoted to an excited state (excitation). The beta particle also loses energy by emitting photons (bremsstrahlung radiation), as it is deflected by the coulomb fields of nuclei. Because the rate of energy loss per unit path length (linear energy transfer, or LET) increases as the velocity of the beta particle slows, a distinct maximum range can be associated with beta particles of known initial energy.

The beta decay energy spectrum for tritium is shown below in Figure E-1. The maximum energy of the tritium beta is 18.591 ± 0.059 keV. The average energy is 5.685 ± 0.008 keV. The maximum range[†] of the tritium beta is 0.58 mg/cm^2.

The absorption of energy from beta particles that emanate from a point source of tritium has been shown to occur nearly exponentially with distance. This is a result of the shape of the beta spectrum as it is subdivided into ranges that correspond with subgroups of initial kinetic energies. As a consequence, the fraction of energy absorbed, F, can be expressed as

$$F = 1 - e^{-(\mu/\rho)(\rho)(x)}, \qquad (E.3)$$

where μ/ρ is the mass attenuation coefficient of the surrounding material, ρ is the density of the surrounding material, and x is the thickness of the surrounding material. For incremental energy absorption calculations, Equation (E.3) can be restated as

$$F = 1 - e^{-\mu x}, \qquad (E.3a)$$

where μ (i.e., the linear attenuation coefficient) is the product of the mass attenuation coefficient (μ/ρ) and the density (ρ), and x is the incremental thickness of choice. In gases at 25° C, at atmospheric pressure, for example, the linear attenuation coefficients for the gases hydrogen (H_2), nitrogen (N_2), and argon (Ar), are 1.81 cm^{-1}, 11.0 cm^{-1}, and 12.9 cm^{-1}, respectively. A 5-mm thickness of air will absorb 99.6 percent of tritium betas. A comparable thickness of hydrogen (or tritium) gas will absorb only 60 percent of the tritium betas.

Absorption coefficients for other media can be estimated by applying correction factors to the relative stopping power (the scattering probability) of the material of interest. For the most part, these will be directly proportional to ratios of electron densities. Examples of tritium beta ranges are shown below in Table E-2. The values shown for tritium gas and for air are stated as STP values.

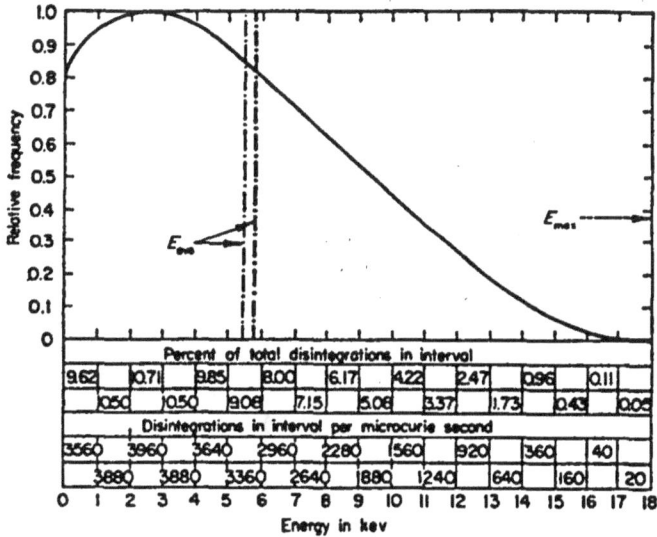

Figure E-1.   Tritium Beta-Decay Energy Spectrum.

[†]   To be technically correct, the term *range* should have the units of distance. In many cases, however, it is more convenient to express the *maximum range* of a particle in terms of the mass per unit area of the absorber needed to stop the particle (with units of mg/cm^2), which is equal to the product of the absorber's density (in units of mg/cm^3) and the range (in units of cm). An advantage of expressing ranges in this way is that, as a practical matter, the masses and areas of thin foils, which are often used in range experiments, are easier to measure than their thicknesses.

Table E-2.
Approximate Ranges of Tritium Betas.

Material	Beta Energy	Range
Tritium Gas	Average	0.26 cm
Tritium Gas	Maximum	3.2 cm
Air	Average	0.04 cm
Water (Liquid)	Average	0.42 μm
Water (Liquid)	Maximum	5.2 μm
Stainless Steel	Average	0.06 μm

E.3.3   Photon Emissions

No nuclear electromagnetic emissions (gamma emissions) are involved in the decay scheme for tritium, although it is worth noting that tritium does produce bremsstrahlung (braking radiation) as its beta particles are decelerated through interactions with nearby matter. For purposes of this document, however, the production of tritium bremsstrahlung radiation can be ignored.

E.4   The Chemical Properties of Tritium

E.4.1   Generic

Although the chemical properties of tritium have been described in great detail, three distinct types of chemical reactions, and one underlying principle in particular, are worth noting here. The reaction types are solubility reactions, exchange reactions, and radiolysis reactions. The underlying principle is Le Chatelier's Principle. An overview of these types of reactions and of Le Chatelier's Principle is presented below.

E.4.2   Solubility Reactions

Elemental hydrogen, regardless of its molecular form (i.e., H_2, HD, D_2, HT, DT, and/or T_2), can be expected to be soluble, to some extent, in virtually all materials. On the atomic or molecular scale, hydrogen-like atoms, diatomic hydrogen-like species, or larger, hydrogen-like-bearing molecules tend to dissolve interstitially (i.e., they diffuse into the crystalline structure, locating themselves inside the normal lattice work of the internal structure). Schematically, such reactions can easily be described in terms of the generic reactions:

$$H_2 + \text{Material} \rightarrow 2H\square\text{Material}, \tag{E.4a}$$

$$^x_1H + \text{Material} \rightarrow {}^x_1H\square\text{Material}, \tag{E.4b}$$

and

$$^3_1H + \text{Material} \rightarrow {}^3_1H\square\text{Material}. \tag{E.4c}$$

Theoretically, however, the underlying mechanics are much more complex. For example, of the generic reactions shown above, none are shown as being reversible. From a chemical perspective, none of these reactions is technically correct because, in most dissolution reactions, the solute that goes in can be expected to be the same solute that comes out. From an operational standpoint, however, experience has shown that, regardless of the tritiated compound that enters into the reaction, an HTO (i.e., a tritiated water vapor) component can be expected to come out. Presumably, this is due to catalytic effects and/or exchange effects that derive from the outward migration of the tritiated species through the molecular layers of water vapor that are bound to the downstream surface of the material.

E.4.3 Exchange Reactions

Driven primarily by isotope effects, exchange reactions involving tritium can be expected to occur at a relatively rapid pace. Moreover, the speed at which reactions of this type can occur can be further enhanced by the addition of energy from radioactive decay. For tritium, therefore, reactions similar to the following can be expected, and they can be expected to reach equilibrium in time frames that range from seconds to hours:

$$CH_4 + 2T_2 \ \Box \quad CT_4 + 2H_2, \qquad (E.5)$$

and

$$2H_2O + T_2 \ \Box \quad 2HTO + H_2. \qquad (E.6)$$

Equation (E.5) describes the preferential form of tritium, as it exists in nature, in the earth's upper atmosphere. Equation (E.6) describes the preferential form of tritium, as it exists in nature, in the earth's lower atmosphere, i.e., in a terrestrial environment.

Equation (E.6) is particularly important because it describes the formation of tritiated water vapor (i.e., HTO) without the involvement of free oxygen (i.e., with no free O_2). A comparable reaction that would involve free oxygen would take the form of a classic inorganic chemical reaction, such as

$$H_2 + T_2 + O_2 \rightarrow 2HTO. \qquad (E.7)$$

But, because a classic inorganic chemical reaction like that depicted in Equation (E.7) can be expected to reach equilibrium in a time frame that ranges from many hours to several days under the conditions normally found in nature, classic inorganic chemical reactions of this type are not necessary for this discussion.

E.4.4 Radiolysis Reactions

It was noted previously in Section E.3.2 that the range of the tritium beta is very short. As a consequence, it follows that virtually all of the energy involved in tritium beta decay will be deposited in the immediate vicinity of the atoms undergoing decay. When the medium surrounding the decaying atoms is tritium gas, tritiated water, or tritiated water vapor in equilibrium with its isotopic counterparts, reactions such as those presented in Equations (E.8) and/or (E.9) below can be expected to dominate. When the medium surrounding the decaying atoms is not a medium that would normally be expected to contain tritium, however, an entire spectrum of radiolysis reactions can be expected to occur.

For typical, day-to-day operations, the most common type of radiolysis reactions in the tritium community can be expected to occur at the interface between the air above a tritium contaminated surface and the tritium contaminated surface itself. For these types of reactions, some of the energy involved in the tritium decay process can be expected to convert the nitrogen and oxygen components in the air immediately above the surface (i.e., the individual N_2 and O_2 components in the air) into the basic generic oxides of nitrogen, such as nitric oxide, nitrous oxide, and nitrogen peroxide (i.e., NO, N_2O, and NO_2, respectively). As the energy deposition process continues, it can also be expected that these simpler oxides will be converted into more complex oxides, such as nitrites and nitrates (i.e., NO_2s and NO_3s, respectively). Because all nitrite and nitrate compounds are readily soluble in water (and/or water vapor), it can further be expected that a relatively large percentage of the available nitrites and nitrates in the overpressure gases will be adsorbed into the mono-molecular layers of water vapor that are actually part of the surface. (See Section E.6, below.) With the available nitrites and nitrates now an integral part of the mono-molecular layers of water vapor, it can finally be expected that the most common type of radiolysis-driven reactions should result in the gradual, low-level build-up of tritiated nitrous and nitric acids on the surfaces of most tritium contaminated materials.

For the most part, this particular type of reaction sequence does not normally present itself as a problem in day-to-day tritium operations because (1) the overall production efficiency for these types of reactions is relatively low, and (2) the materials used for the construction of most tritium-handling systems are not susceptible to attack by nitrous and/or nitric acids. By contrast, however, it should be noted that other types of radiolysis-driven reactions can be expected to occur with tritium in the presence of compounds containing chlorides and/or fluorides, and that these can easily lead to chloride/fluoride induced stress-corrosion cracking. (See, for example, the discussion on *Materials Compatibility Issues*, in Section E.7, below.)

One additional point that is worth noting about radiolysis-driven reactions is that their long-term potential for causing damage should not be underestimated. Although the overall production efficiency for these types of reactions might be expected to be relatively low, the generation of products from these types of reactions can, on the other hand, be expected to occur continuously over relatively long periods of time (i.e., 10–20 years, or more). As a consequence, the long-term effects from these types of reactions can be difficult to predict, especially because very little is known about the long-term, synergistic effects of low-level, tritium micro-chemistry. (See Sections E.7 and E.8, below.)

E.5   Le Chatelier's Principle

A chemical restatement of Newton's Third Law of Motion, Le Chatelier's Principle states that when a system at equilibrium is subjected to a perturbation, the response will be such that the system eliminates the perturbation by establishing a new equilibrium. When applied to situations like those depicted in Equations (E.5) and (E.6) above, Le Chatelier's Principle states that, when the background tritium levels are increased in nature (by atmospheric testing, for example), the reactions will be shifted to the right in order to adjust to the new equilibrium conditions by readjusting to the naturally occurring isotopic ratios. Thus, we get reactions of the type

$$CH_4 + 2T_2 \leftrightarrow CT_4 + 2H_2, \tag{E.5a}$$

and

$$2H_2O + T_2 \leftrightarrow 2HTO + H_2. \tag{E.6a}$$

The inverse situation also applies in that, when the background tritium levels are decreased in nature, the reactions will be shifted back to the left, by again readjusting to the naturally occurring isotopic ratios, i.e.,

$$CH_4 + 2T_2 \leftrightarrow CT_4 + 2H_2, \tag{E.5b}$$

and

$$2H_2O + T_2 \leftrightarrow 2HTO + H_2. \tag{E.6b}$$

By itself, Le Chatelier's Principle is a very powerful tool. When applied singularly, or to a sequential set of reactions like those depicted in Equations (E.5), (E.5a), and (E.5b), and/or (E.6), (E.6a), and (E.6b), Le Chatelier's Principle shows that exchange reactions of the types depicted above tend to behave as springs, constantly flexing back-and-forth, constantly readjusting to changing energy requirements, in a constantly changing attempt to react to a new set of equilibrium conditions.

Since elemental hydrogen, regardless of its molecular form (i.e., H_2, HD, D_2, HT, DT, and/or T_2), can be expected to dissolve to some extent in virtually all materials, Le Chatelier's Principle can be expected to work equally as well on solubility reactions, like those shown above in the generic Equations (E.4a), (E.4b), and (E.4c). These will be covered in more detail under the heading of *Bulk Contamination Modeling*. (See Section E.6.4, below.)

E.6 Modeling the Behavior of Tritium

Any model of the behavior of tritium starts with the assumption that all three hydrogen isotopes coexist in nature, in equilibrium with each other, in the nominal isotopic ratios described above in Table E-1. To this is added the consequences predicted by Le Chatelier's Principle. From both, we get the fundamental relationship,

$$2H_2 + 2D_2 + 2T_2 \rightleftharpoons H_2 + HD + D_2 + HT + DT + T_2. \qquad (E.8)$$

In a terrestrial environment, virtually all of the tritium that exists in nature exists as water or water vapor. Correcting this situation for the natural conversion to water and/or water vapor, Equation (E.8) becomes

$$2H_2O + 2D_2O + 2T_2O \rightleftharpoons H_2O + HDO + D_2O + HTO + DTO + T_2O. \qquad (E.9)$$

It can also be assumed that the surfaces of all terrestrially bound objects are coated with a series of monomolecular layers of water vapor. In the final step, it can be assumed that the innermost layers of water vapor are very tightly bound to the actual surface, that the intermediate layers of water vapor are relatively tightly to relatively loosely bound, and that the outermost layers of water vapor are very loosely bound. (See Figure E-2.)

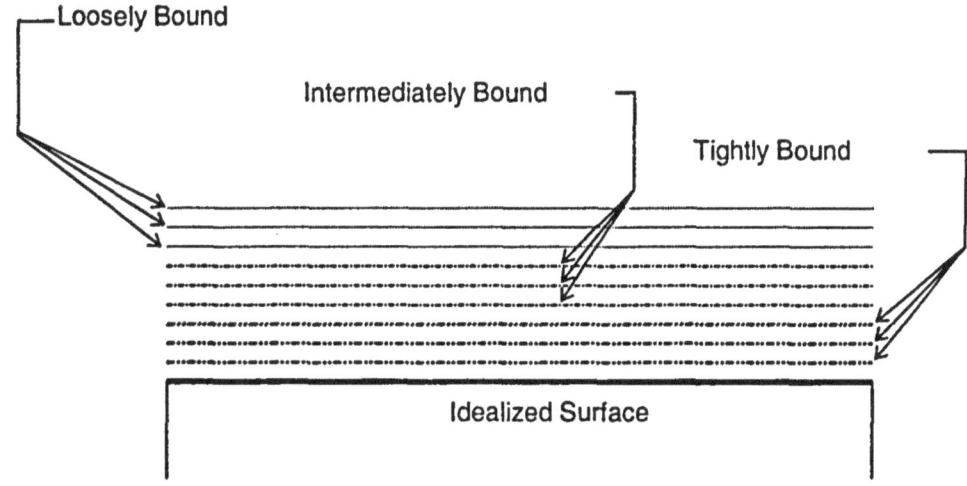

Figure E-2.   Idealized Surface Showing Idealized Monomolecular Layers of Water Vapor.

E.6.1   Surface Contamination Modeling

When an overpressure of tritium is added to the system (i.e., the surface, in this case), a perturbation is added to the system, and Le Chatelier's Principle tells us that the tritium levels in the monomolecular layers of water will be shifted to the right, i.e.,

$$2H_2O + 2D_2O + 2T_2O \rightleftharpoons H_2O + HDO + D_2O + HTO + DTO + T_2O. \qquad (E.9a)$$

Tritium is incorporated first into the loosely bound, outer layers, then into the intermediate layers, and finally into the very tightly bound, near surface layers. When the overpressure is removed, the system experiences a new perturbation. In this case, however, the perturbation is in the negative direction, and the system becomes the entity that contains the excess tritium. Le Chatelier's Principle, in this case, indicates that the tritium levels in the monomolecular layers of water will be shifted back to the left, i.e.,

$$2H_2O + 2D_2O + 2T_2O \rightleftharpoons H_2O + HDO + D_2O + HTO + DTO + T_2O. \qquad (E.9b)$$

The tritium that had previously been incorporated into the monomolecular layers now begins to move out of the layers, in an attempt to return to background levels.

The movement of tritium into the monomolecular layers of water vapor is generically referred to as "plate-out." The movement of tritium out of the monomolecular layers of water vapor is generically referred to as "outgassing."

E.6.2   Plate-Out Expectations

When the concentration gradients have been small and/or the exposure times have been short, only the outermost, loosely bound, monomolecular layers of water vapor will be affected. Under such circumstances, the surface contamination levels will range from no detectable activity to very low levels; that is, up to a few tens of disintegrations per minute per 100 square centimeters (dpm/100 cm^2). Since only the outermost monomolecular layers are affected, and since these layers are easily removed by a simple wiping, the mechanical efforts expended to perform decontamination on such surfaces will, if any, be minimal.

When the concentration gradients have been relatively large and/or the exposure times have been relatively long, the affected monomolecular layers will range down into the intermediately bound layers (i.e., the relatively tightly to relatively loosely bound layers). Under these circumstances, the surface contamination levels will range from relatively low to relatively high (i.e., from a few hundred to a few thousand dpm/100 cm^2). Because the tritium has now penetrated beyond those levels that would normally be easily removed, the mechanical efforts expended to decontaminate such surfaces will be more difficult than those described above.

When the concentration gradients have been large and/or the exposure times have been long, the affected monomolecular layers will range all the way down into the very tightly bound layers. The tritium will have penetrated down into the actual surface of the material, itself; see "Bulk Contamination Modeling," below. Under such circumstances, the surface contamination will range from relatively high to very high levels (i.e., from a few tens of thousand to several hundred thousand dpm/100 cm^2), and the mechanical efforts expended to decontaminate such surfaces could be very difficult.

E.6.3   Outgassing Expectations

The phenomenon of outgassing is rarely a problem under the first of the exposure situations described above, i.e., situations in which the concentration gradients have been small and/or the exposure times have been short. However, when systems that have been exposed to even small amounts of tritium for long-to-very-long periods of time are suddenly introduced to room air, or any sudden change in its equilibrium situation, Reactions (E.5), (E.5a), and (E.5b), Reactions (E.6), (E.6a), and (E.6b), and Reactions (E.9), (E.9a), and (E.9b) can be thought of as *springs*, and the initial phenomenon of outgassing can be described as damped harmonic motion. Under such circumstances, therefore, a relatively large, initial "puff" of HTO will be released from the monomolecular layers of water vapor, followed by a relatively long, much smaller trailing release. Because several curies of HTO can be released in a few seconds, and several tens of curies can be released in a few minutes, the speed of the "puff" portion of the release should not be underestimated. The duration of the trailing portion of the release should not be underestimated either. Depending on the concentration gradients involved and/or the time frames involved in the plate-out portion of the exposure, the trailing portion of the release can easily last from several days to several months or even years.

As the trailing portion of the release asymptotically approaches zero, the outgassing part of the release becomes too small to measure on a real-time basis, and the tritium levels involved in any given release can only be measured by surface contamination measurement techniques. Under such circumstances, the situation reverts back to the circumstances described above under the heading of "Plate-Out Expectations." With no additional influx of tritium, tritium incorporated into all of the monomolecular layers of water vapor will eventually return to background levels, without human intervention, regardless of the method or level of contamination.

E-8

E.6.4 Bulk Contamination Modeling

When an overpressure of tritium is added to the system (i.e., the surface of an idealized material), Le Chatelier's Principle indicates that the tritium levels in the mono-molecular layers of water will be shifted to the right; i.e.,

$$2H_2O + 2D_2O + 2T_2O \rightleftharpoons H_2O + HDO + D_2O + HTO + DTO + T_2O. \tag{E.9a}$$

Tritium is incorporated first into the loosely bound, outer layers, then into the intermediate layers, and finally into the very tightly bound, near-surface layers. As the tritium loading in the near-surface layers builds, the disassociation processes that proceed normally as a result of the tritium decay make an overpressure of tritium available in the atomic form (i.e., as T). Relative to the normal amounts of elemental hydrogen that can be expected to be dissolved in the material, the availability of excess tritium in the atomic form represents a different type of perturbation on a system, and the available tritium begins to dissolve into the actual surface of the bulk material. As the local saturation sites in the actual surface of the bulk material begin to fill, the tritium dissolved in the surface begins to diffuse into the body of the bulk material. At that point, the behavior of the tritium in the body of the bulk material becomes totally dependent on the material in question.

E.7   Materials Compatibility Issues

Elemental hydrogen, regardless of its form (H_2, D_2, T_2, and all combinations thereof), can be expected to dissolve to some extent in virtually all materials. For simple solubility reactions, such as

$$H_2 + \text{Material} \rightarrow 2H \cdot \text{Material}, \tag{E.4a}$$

$$^x_1H + \text{Material} \rightarrow {^x_1}H \cdot \text{Material}, \tag{E.4b}$$

and

$$^3_1H + \text{Material} \rightarrow {^3_1}H \cdot \text{Material}, \tag{E.4c}$$

basic compatibility issues should be considered. As a general rule, the solubility of tritium in pure metals and/or ceramics should have a minimal effect, at normal room temperatures and pressures, except for the possibility of hydrogen embrittlement. For alloyed metals, such as stainless steel, similar considerations apply, again, at normal room temperatures and pressures. For alloyed metals, however, additional consideration must be given to the possible leaching of impurities from the alloyed metal, even at normal room temperatures and pressures. [In LP-50 containment vessels, for example, the formation of relatively large amounts of tritiated methane (i.e., up to 0.75 mole percent of CT_4) has been noted after containers of high purity tritium have been left undisturbed for several years. The formation of the tritiated methane, in this case, has long been attributed to the leaching of carbon from the body of the stainless steel containment vessel.]

E.7.1   Pressure Considerations

Under increased pressures (e.g., from a few tens to several hundred atmospheres), however, the general rules no longer apply for, in addition to the possibility of hydrogen embrittlement and possible leaching effects, helium embrittlement is also possible. Helium embrittlement tends to occur as a result of the dissolved tritium decaying within the body of the material, the resultant migration of the helium-3 atoms to the grain boundaries of the material, the localized agglomerations of the helium-3 atoms at the grain boundaries, and the resultant high-pressure build-ups at these localized agglomerations.

E.7.2   Temperature Considerations

Under increased temperature situations, the matrix of solubility considerations becomes even more complicated because virtually all solubility reactions are exponentially dependent on temperature. In the case of diffusional flow through the walls of a containment vessel, for example, it can be assumed that steady-state permeation will have been reached when

$$\left(\frac{Dt}{L^2}\right) \cong 0.45, \tag{E.10}$$

where D = the diffusion rate in cm²/s, t = the time in seconds, and L = the thickness of the diffusion barrier. For type 316 stainless steel, the value for the diffusion rate is

$$D = 4.7 \times 10^{-3} \, e^{-12,900/RT}, \tag{E.10a}$$

and the corresponding value for R, in the appropriate units, is 1.987 cal/mole K. With a nominal wall thickness of 0.125 inches (i.e., 0.318 cm), Equation (E.10) indicates that it will take about 875 years to reach steady-state permeation, at a temperature of 25° C. At 100° C, the time frame will be reduced to about 11 years, and at 500° C, it only takes about 12 hours.

E.8 Organics

With the introduction of organic materials into any tritium handling system, the matrix of solubility considerations becomes complicated to its maximum extent because the simple solubility reactions, such as those shown above as Equations (E.4a), (E.4b), and (E.4c), are no longer working by themselves. With the availability of free tritium dissolved into the internal volume of the organic material, the molecular surroundings of the organic material see a local perturbation in their own internal systems, and Le Chatelier's Principle indicates that the system will adjust to the perturbation with the establishment of a new equilibrium. Under such circumstances, exchange reactions can be expected to dominate over simple solubility reactions, and the available tritium can be expected to replace the available protium in any—and all—available sites. Once the tritium has been incorporated into the structure of the organic material, the structure begins to break down from the inside out, primarily as a result of the tritium decay energy.

The specific activity of tritium gas at atmospheric pressure and 25° C is 2.372 Ci/cm³. The expected range of the average energy tritium beta particle in unit density material is only 0.42 μm. This means that all energy from the decay of the dissolved tritium is deposited directly into the surrounding material. At 2.372 Ci/cm³, this becomes equivalent to 2.88×10^4 rads/hr.

The general rule for elastomers used for sealing is that total radiation levels of 10^7 rads represent the warning point that elastomers may be losing their ability to maintain a seal. At 10^8 rads, virtually all elastomers used for sealing lose their ability to maintain a seal. Typical failures occur as a result of compression set (i.e., the elastomer becomes brittle and loses its ability to spring back). At 10^6 rads, on the other hand, the total damage is relatively minor, and most elastomers maintain their ability to maintain a seal. At 10^7 rads, the ability of an elastomer to maintain a seal becomes totally dependent on the chemical compounding of the elastomer in question. It only takes about 2 weeks for an elastomer to receive 10^7 rads at a dose rate of 2.88×10^4 rads/hr. Elastomers, therefore, cannot be used for sealing where they might be exposed to high concentrations of tritium.

Similar analogies can be drawn for all organic materials. The preferred rule of thumb is the use of all organic materials should be discouraged wherever they might be exposed to tritium. Since this is neither possible nor practical, the relative radiation resistance for several elastomers, thermoplastic resins, thermosetting resins, and base oils is shown graphically in Figure E-3, Figure E-4, Figure E-5, and Figure E-6, respectively.

The damage done to organic materials by the presence of tritium in the internal structure of the material is not limited to the more obvious radiation damage effects. Tritium, particularly in the form of T⁺, has the insidious ability to leach impurities (and non-impurities) out of the body of the parent material. In many cases, particularly where halogens are involved, the damage done by secondary effects, such as leaching, can be more destructive than the immediate effects caused by the radiation damage. In one such case, the tritium contamination normally present in heavy water up to several curies per liter was able to leach substantial amounts of chlorides out of the bodies of neoprene‡ O-rings that were used for the seals. The chlorides leached out of the O-rings were subsequently deposited into the stainless steel sealing surfaces above and below the trapped O-rings, which led directly to the introduction of chloride-induced, stress-corrosion cracking in the stainless steel.

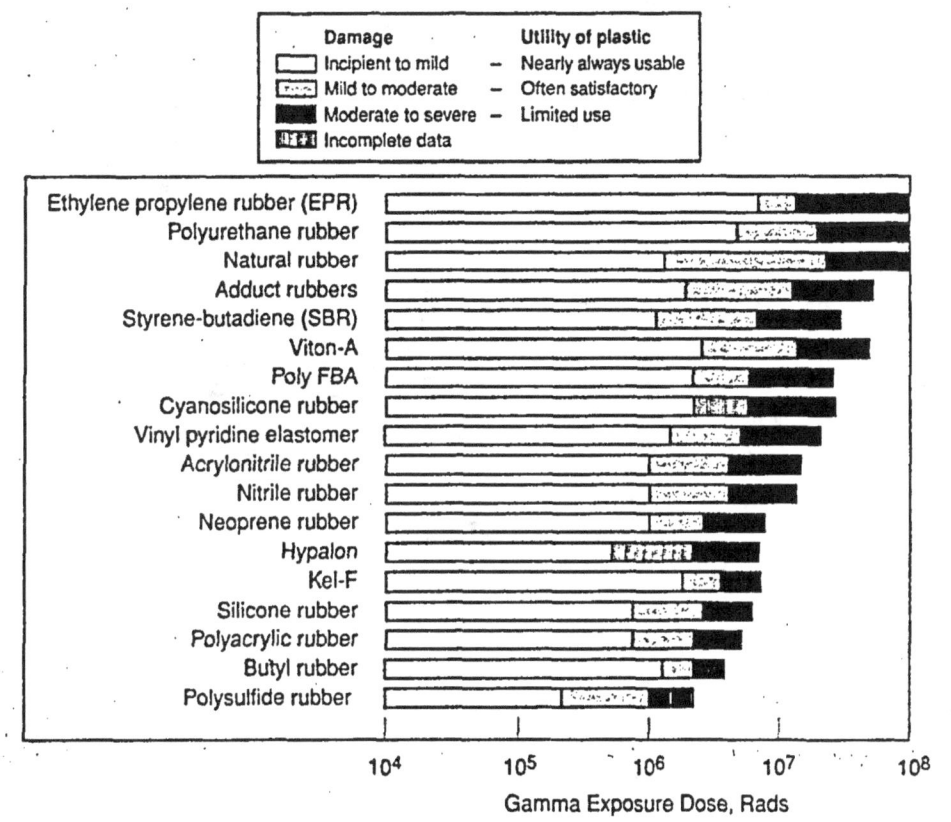

Figure E-3.   Relative Radiation Resistance of Elastomers.

The operational conditions that set up the introduction of the stress-corrosion cracking were moderately elevated temperatures (i.e., less than 100° C), low pressures (i.e., less than 3 atmospheres), and exposure times of 3-5 years. Fortunately, the damage was discovered before any failures occurred. The neoprene O-rings were removed, and the seal design was changed to a non-O-ring type of seal.

In a second case, six failures out of six tests occurred when high-quality Type 316 stainless steel was exposed to tritium gas in the presence of Teflon™ shavings and 500-ppm moisture. All of the failures were catastrophic, and all were the result of massively induced stress-corrosion cracking. The conditions that set up the introduction of the massively induced stress-corrosion cracking in this case were moderately elevated temperatures (i.e., 104° C), relatively high pressures (i.e., 10,000 to 20,000 psi), and exposure times that ranged from 11 to 36 hours. Since the time to failure for all the tests was directly proportional to the pressure (i.e., the higher pressure tests failed more quickly than the lower pressure tests), since identical control tests with deuterium produced no failures, and since comparable testing without the Teflon™ shavings indicated no failures after 3,200 hours, it was concluded that fluorides were being leached out of the Teflon™ and deposited directly into the bodies of the stainless steel test vessels. An interesting sideline to this test is that, after the tests, the Teflon™ shavings showed no obvious signs of radiation damage (i.e., no apparent discoloration or other change from the original condition).

---

‡   The proper chemical name for neoprene is "chlorobutadiene."

E-11

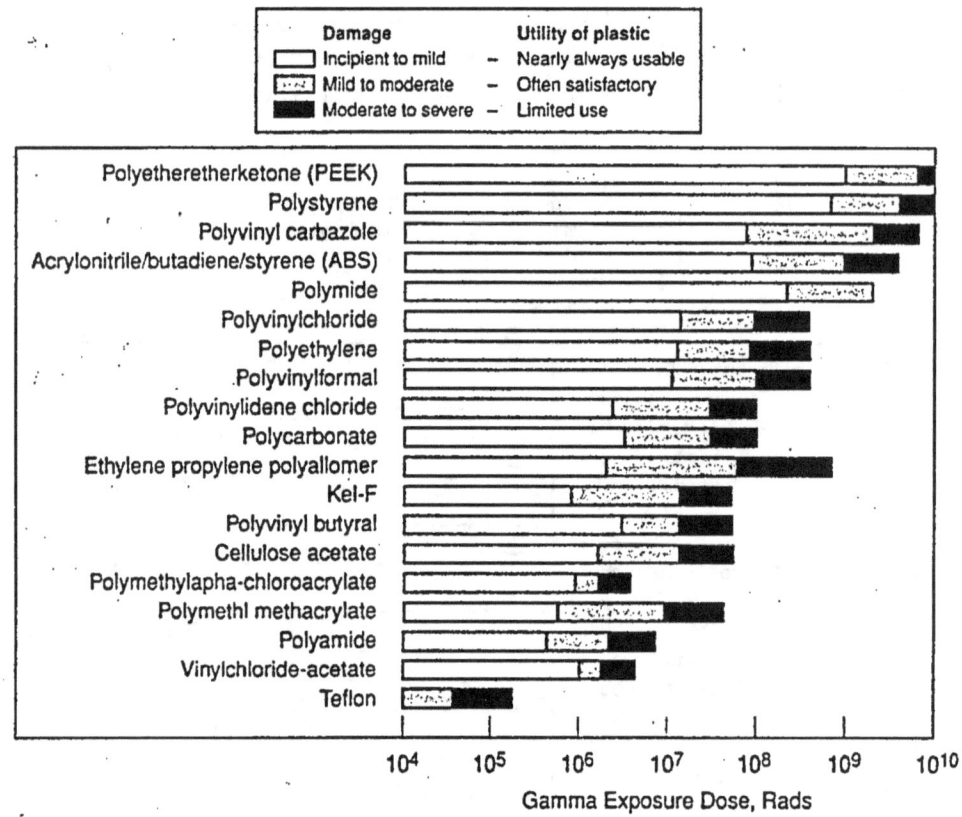

Figure E-4.   Relative Radiation Resistance of Thermoplastic Resins.

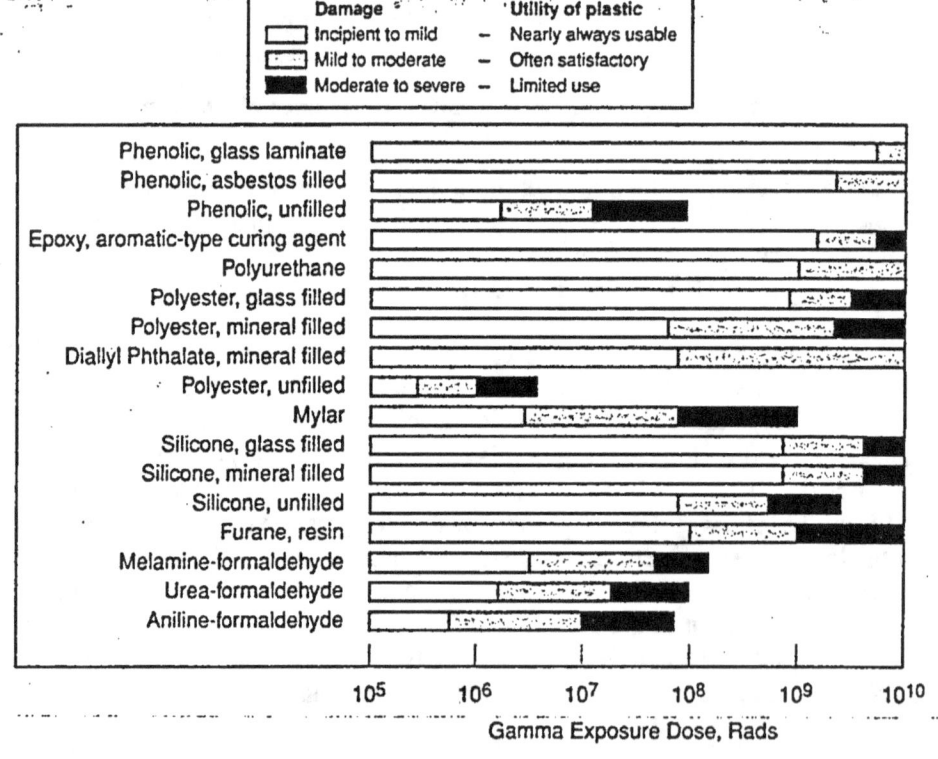

Figure E-5.   Relative Radiation Resistance of Thermosetting Resins.

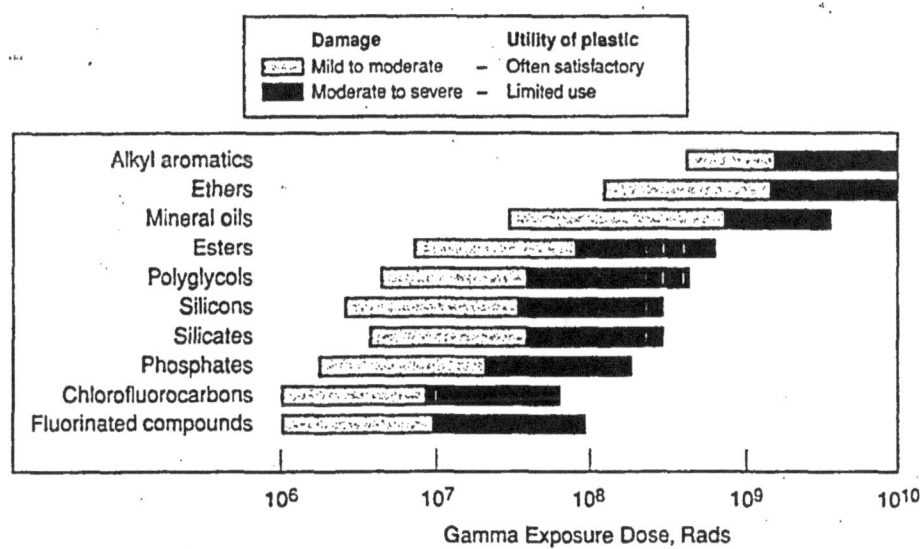

Figure E-6. Relative Radiation Resistance of Base Oils.

E.9 Outgassing from Bulk Materials

Discussions on the outgassing from bulk materials can be subdivided into two parts: 1) outgassing from surfaces that have been wetted with tritium, and 2) outgassing from surfaces that have not been wetted with tritium. For surfaces that have been wetted with tritium, the behavior of the outgassing should be virtually identical to that described above. For surfaces that have not been wetted with tritium, it should be assumed that the source of the outgassing is from tritium that has been dissolved in the body of the parent material.

As the saturation level in the body of the bulk material is reached, the dissolved tritium begins to emerge from the unexposed side of the material surface, where it then begins to move through the monomolecular layers of water vapor on that side. In the initial stages, the pattern of the tritium moving into these monomolecular layers tends to resemble the reverse of that described in the surface contamination model described above (i.e., the tritium is incorporated first into the very tightly bound, near-surface layers, then into the intermediate layers, and finally into the loosely bound, outer layers). As the tritium saturation levels in the body of the bulk material gradually reach steady-state, the tritium levels moving into the monomolecular layers gradually build over time, and the pattern slowly changes from one of a reverse surface contamination model to one of a reverse outgassing model (i.e., the level of outgassing from any given surface can be expected to increase until it too reaches a steady-state, equilibrium level with its own local environment).

E.10 References

U.S. Department of Energy, *Design Considerations*, DOE-HDBK-1132-99, U.S. Government Printing Office, Washington, D.C., April 1999. (Note: The bulk of the material presented in the sections has been adapted from this reference. See, in particular, Sections 2.10.1 through 2.10.6, pp. 1–86 through 1–109.)

E.11 Suggested Additional Reading

U.S. Department of Energy, *Primer on Tritium Safe Handling Practices*, DOE-HDBK-1079-94, December 2001.

U.S. Department of Energy, *Radiological Training for Tritium Facilities*, DOE-HDBK-1105-96, December 2001.

U.S. Department of Energy, *Tritium Handling and Safe Storage*, DOE-HDBK-1129-99, March 1999.

Appendix F: Biological Properties of Tritium and Tritium Health Physics

(Note: With the exception of Sections F.5.1.1.1, F.5.1.1.2, and F.5.1.1.3, the bulk of the material presented below was adapted from Sections 3 and 4 of the Department of Energy's, "Health Physics Manual of Good Practices for Tritium Facilities," published in 1991.[F-1] Although some of the information may appear to be somewhat dated, the basic concepts behind the information have not changed since that time.* See also the information presented in Appendix E.)

F.1   Biological Properties of Tritium

F.1.1   General

Tritium is usually encountered in the workplace as tritium gas (HT, DT, or T_2) or as tritiated water, or water vapor (i.e., HTO, DTO, or T_2O). Other forms of tritium also exist, such as tritiated surfaces, metal tritides, tritiated pump oil, and tritiated gases. While some minor isotopic differences in reaction rates have been noted, deuterated and tritiated compounds generally have the same biological properties as the hydrogenated compounds. These various tritiated compounds will have a wide range of uptake and retention in humans under identical exposure conditions. Tritium gas, for example, represents one end of the spectrum, in that the body has no physiological use for elemental hydrogen regardless of its isotopic form and can easily be exhaled. Water vapor, on the other hand, represents the opposite end of the spectrum because it is readily taken up and retained by the body. Less is known about the uptake and retention of other tritiated compounds.

F.1.2   The Metabolism of Gaseous Tritium

The biological mechanisms for inhalation exposure to gaseous tritium are similar to the biological mechanisms for airborne nitrogen: 1) small amounts of the gas will be dissolved in the bloodstream according to the laws of partial pressures; 2) the dissolved gas will be circulated in the bloodstream with a resident half-time of about 2 minutes; and 3) most of the gas will subsequently be exhaled along with the gaseous waste products carbon dioxide and normal water vapor. A small percentage of the gaseous tritium will be converted to the oxide form (HTO), most likely in the gastrointestinal tract. Early experiments showed that the total biological conversion to HTO can range from 0.004% to 0.1% of the total gaseous tritium inhaled. More recent experiments with six volunteers resulted in a conversion of 0.005% with an uncertainty in the average conversion rate of ± 0.0008%.

Skin absorption of gaseous tritium has been found to be negligible when compared to inhalation. Small amounts of tritium can enter skin through contact with contaminated surfaces and result in elevated organically bound tritium in tissues and in urine. (See Sections F.1.4 and F.1.5, below.) Hence, for gaseous tritium exposures, there is a lung dose from the tritium in the air in the lung, and a whole body dose from the tritium gas that has been converted to water. This in vivo converted tritiated water will of course act like an exposure to tritiated water.

F.1.3   The Metabolism of Tritiated Water

The biological incorporation (uptake) of airborne HTO can be extremely efficient—up to 99% of inhaled HTO can be taken into the body within seconds. Ingested liquid HTO is almost completely absorbed by the gastrointestinal tract and quickly appears in the venous blood. Within minutes, it can be found in varying concentrations in the various organs, fluids, and tissues of the body. Skin absorption mechanisms also become important because the internal temperature of the body is regulated, to a large extent, by "breathing" water vapor in and out through the pores of the skin. For skin temperatures in the range of 30 to 40 °C, it has been shown that

---

*   Additional Note: Because the bulk of the information presented in this Appendix is presented in a paraphrased format, it is suggested that the reader refer directly to Reference F-1 for additional information, which does include all of the references to the original citations.

the percutaneous absorption of HTO is about equal to that for HTO by inhalation. Thus, it can be expected that, independent of the absorption mechanism, absorbed HTO will be uniformly distributed in all biological fluids in time frames that range from 45 minutes to 2 hours. Therefore, very shortly after an exposure to HTO, the tritium will be uniformly spread throughout the tissue of the body in body water and in the exchangeable (labile) hydrogen sites in organic molecules. This tritium will have a retention that is characteristic of water. A small fraction of the tritium will become incorporated into non-exchangeable hydrogen sites in organic molecules giving rise to a long-term retention that is characteristic of the turn over of cellular components, which can be adequately modeled as the sum of two exponentials. Hence, retention of tritiated water can be described as the sum of two exponentials, one characteristic of body water, and two longer-term components that represent tritium incorporated into non-labile cellular hydrogen sites.

F.1.4   The Metabolism of Other Tritiated Species

As mentioned above, most tritium will be in the form of tritiated hydrogen gas or tritiated water. However, tritium handling operations will result in the production of other forms of tritium, such as tritiated surfaces, metal tritides, pump oils, and a wide variety of "other" tritiated species, some of which are discussed below.

F.1.4.1 Tritiated Surfaces

Studies have shown that when there is contact between skin and a surface that has been exposed to high concentrations of tritium gas, tritium is transferred to the body in an organic form. This organically bound tritium gives rise to elevated tritium concentrations in skin at the point of contact and in other tissues, and a large amount of organically bound tritium in urine. The full metabolic pathway of this organically bound tritium is unknown, but models that have been developed suggest that the dose to skin at the point of contact is the limiting factor in exposures of this type.

F.1.4.2 Metallic Tritides

Although a broad spectrum of metals are commonly used for the storage, pumping, and packaging of tritium, there is little data on their metabolic properties. However, some compounds are unstable in air, e.g., uranium tritide and lithium tritide. For these, exposure to air produces totally different results: uranium tritide, being pyrophoric, releases large quantities of tritiated water; lithium tritide, being a hydroxyl scavenger, releases large quantities of tritium gas.

At the other end of the spectrum, metallic tritides such as titanium, niobium, and zirconium tritides are very stable in air. For these, the organ of concern must be primarily the lung, and one relies on lung deposition models such as the one presented in the International Commission on Radiation Protection's Publication 30 (ICRP-30).[F-2] However, there are difficulties with using such models. Depending on the particle size distribution of the metallic tritide inhaled, lung retention estimates can be in error by up to 80 percent. Also, cross-correlations of lung retention estimates are based on the tritium leaching ability of biological fluids, which are dependent on the chemical and physical form of the material in question. These particles may also produce organically bound tritium from contact with lung tissue, and this would further compound the metabolic uncertainties.

F.1.4.3 Generic Tritiated Solids

The formation of generic tritiated solids can be expected to occur in all normal solid materials that are routinely exposed to tritium. Depending on the composition of the material, tritiation will occur through exchange reactions and/or through mechanisms such as solubility, permeation, and diffusivity. The specific activity of such materials can be expected to vary in relation to the relative concentration of the exposing gas, the relative humidity of the exposing gas, and the total reaction time. Radiation damage may also be expected, particularly in cases where possible exposure mechanisms lead to embrittlement.

Because little is known about the metabolic behavior of generic tritiated solids, each must be considered separately. For example, solid materials that tend to become embrittled should be considered in the same metabolic category as metallic tritides. Such materials would include, but not be limited to, Teflon® valve seats (from dry environs). Other materials, such as those that degrade over time or those that give up their tritium easily (outgas), can be considered as possible inhalation hazards, possible skin absorption hazards, or both.

F.1.4.4 Tritiated Liquids

Next to HTO, the most commonly encountered tritiated liquid is tritiated vacuum pump oil. Comparisons between facilities have shown that the specific activities of pump oils can easily range from a few mCi/l to a few tens of Ci/l. The wide range in specific activities may be due to situation-specific variations in total throughputs for tritium and ambient water vapor. As a first approximation, the metabolic routes for tritiated vacuum pump oils can be taken as being similar to the metabolic routes for HTO.

Next to pump oils, the most commonly encountered group of tritiated liquids is tritiated solvents. Since all solvents, by their nature, can be expected to have a skin absorption pathway, and since most solvents are relatively volatile, the metabolic pathways for tritiated solvents can, as a first approximation, be expected to be similar to the pathways for HTO. However, families of solvents have specific organs of concern and, in most cases, the initial organ of interest will not be the body water, but the liver. Hence, exposure to tritiated solvents may result in significant differences between the establishment of body water equilibria from that observed for tritiated water.

The error in uptake and retention introduced by treating tritiated liquids as HTO will vary greatly with the individual chemical form.

F.1.4.5 Tritiated Gases

Although few gaseous reactions can compete with the energetically favored formation of HTO, other tritiated gases, such as tritiated methane, can be formed. The details of the metabolic pathways should be generally similar to gaseous tritium. Again, the errors introduced by this approximation are unknown.

F.1.5 Metabolic Elimination

F.1.5.1 Single Compartment Modeling of HTO Retention

Studies of biological elimination rates in humans for heavier-than-normal water species go back to 1934, when the body water turnover rate of a single subject was measured using HDO. Since that time, several additional studies have been conducted on a number of subjects with HDO and HTO, the HTO studies being more prevalent. A summary of these data is presented in Table F-1.

A simple average of the data summarized in Table F-1 suggests a value of 9.4 days for the measured biological half-life. Also, the data deviate from this simple average by as much as ± 50%. As is discussed below, there are good reasons for such large deviations.

As a first approach to modeling the observed biological half-life, one can use the equation,

$$A = A_0 \, e^{-(\ln 2 \, t)/(T_{Bio})}, \tag{F.1}$$

where A_0 is the total body water mass, A is the amount of body water remaining after a given time (t), and T_{Bio} is the biological half-life.

From Reference Man data, i.e., ICRP-25,[F-3] values of 42 kg and 3 kg are obtained for the total body water mass and the average daily throughput of water, respectively. Thus the elimination rate is 3/42 = 0.0714 day^{-1} and the theoretical biological half-life for HTO is

$$T_{Bio} = \ln 2/0.0714 = 9.7 \text{ days}, \tag{F.2}$$

which compares very favorably with the 9.4 day average value determined from Table F-1.

The above modeling and values are also based on the assumption that the biological half-life of tritium will be a function of the average daily throughput of water. This part of the hypothesis, therefore, must also be in agreement with experimental and theoretical crosschecks.

Table F-1   Heavier-than-Normal Biological Half-Life

Water Species	Number of Subjects	Measured T_{Bio} (Days)
HDO	1	9 to 10
HDO	21	9.3 ± 1.5
HTO	8	9 to 14
HTO	20	5 to 11
HTO	8	9.3 to 13
HTO	10	7.5 ± 1.9
HTO	5	9.5 (Average)
HTO	6	8.5 (Average)
HTO	310	9.5 ± 4.1

It has been observed experimentally that, when the water intake was 2.7 *l*/day, the half-life for HTO was 10 days; when the water intake was increased to 12.8 *l*/day, the half-life dropped to 2.4 days. Using these values, Equation F.1 produces values of 10.4 days and 1.9 days for the respective half-lives. Agreement of experimental observations with the simple model is very good, and for the high intake value, the lack of better agreement should not be a serious concern considering model simplicity. Without medical intervention (i.e., diuretics), the metabolic efficiency of the processes of forced fluids can require modification of the model. Other factors that affect the biological half-life of HTO in the human body are discussed below.

Comparisons have also been made of biological half-lives versus mean outdoor temperatures at the time of tritium uptake. The data suggest that biological half-lives are shorter when assimilations occur in the warmer months. For example, the 7.5 ± 1.9 day half-life shown in Table F-1 begins to fall into line when it is noted that the data were taken in Southern Nigeria where the mean outdoor temperature averages 80 °F. In contrast, the 9.5 ± 4.1 day half-life shown in Table F-1 was determined over a multi-year period in North American climes, where the mean outdoor temperature averaged 63 °F. Such findings are consistent with metabolic pathways involving sensible and insensible perspiration. As such, the skin absorption/desorption pathways can become an important part of body metabolic throughput of normal water.

Lifestyles also have significant potential influence on the variation of biological half-lives. In one case, for example, the biological half-life of an adult male was followed for approximately 4 months following an acute exposure during which time the half-life appeared to fluctuate back and forth between 4 and 10 days at regular intervals. Closer scrutiny revealed that the subject was a weekend jogger. As a result, the appearance of two, very different biological half-lives was totally valid.

Variations in biological half-lives have also been shown to be inversely correlated with age. In these cases, however, the data suggest that age correlations introduce variations in the biological half-life of no more than ± 20%. When compared to reduction factors of 50 to 250% produced by total fluid throughput and/or skin temperature correlations, age correlations are a secondary correction.

F.1.5.2 Multi-Compartment Modeling

For single compartment modeling, the half-life of interest is that for HTO in the body water. Although it has been observed that the half-life can vary by more than a factor of two for the same person, the HTO component of the biological half-life can be expected to be about 10 days. As was noted in Section F.1.3, however, prolonged exposures can be expected to show signs of two additional components that range from 21 to 30 days and 250 to 550 days, respectively. The former reflects the existence of labile organic pool; the latter suggests the existence of a more tightly bound organic pool.

For purposes of dose calculations, however, the overall contribution from organically bound tritium has been found to be relatively small, i.e., less than about 5%. The ICRP methods for computing the Annual Limits on Intake in air and water utilize the body water component only, including the assumption of a 10-day biological half-life.[F-2]

F.2   Bioassay and Internal Dosimetry

Exposure to tritium oxide (HTO) is by far the most important type of tritium exposure, and it results in the distribution of HTO throughout the body's soft tissue. The HTO enters the body by inhalation or skin absorption. When immersed in airborne HTO, intake through the lungs is approximately twice that absorbed through the skin. The average biological half-life of tritium is 10 days, but it can vary naturally by 50% or more and is dependent on the body-water turnover rate. This has been verified by calculation and by actual measurements of tritium concentrations in body water following exposure. Following exposure to HT, the gas is taken into the lungs and, according to the laws of partial pressures, some is dissolved in the blood stream, which distributes the HT to the body water.

When a person is exposed to HT in the air, two kinds of exposures actually result: one to the lungs and one to the whole body. According to ICRP-30, the lung exposure is the critical one resulting in an effective dose 25,000 times less than would result from an equal exposure to HTO (for workers doing light work).[†] However, during exposure to HT, a small fraction of the tritium in the blood is transferred to the GI tract where it is rapidly oxidized by enzymes in the gut. This results in a buildup of HTO, which remains in the body (with its usual half-life), while the HT is rapidly eliminated following the end of the exposure. The resultant dose from the exposure to this HTO is roughly comparable to the effective dose from the lung exposure to HT. Thus, for both HTO and HT exposures, a bioassay program that samples body water for HTO is an essential element of a good personnel-monitoring program for tritium.

F.2.1   Sampling Schedule and Technique

Following an exposure to HTO, it is quickly distributed throughout the blood system and, within 1 to 2 hours, throughout the extra- and intra-cellular volumes and the remaining body water. Once equilibrium is thus established, the tritium concentration is found to be the same in samples of blood, sputum, and urine. For bioassay purposes, urine is normally used for determining tritium concentrations in body water.

[†]   As was noted at the beginning of this section, the bulk of the information presented in this section was originally published in 1991. Since that time, more up-to-date dose, and dose assessment, models have been developed. See, for example, the references cited at the end of this section under the heading of *Suggested Additional Reading*, i.e., 1) the EPA's Federal Guidance Report #13, and 2) "Tritium Doses from Chronic Atmospheric Releases: A New Approach Proposed for Regulatory Compliance," both of which were published in 2002.

Workers potentially or casually exposed to tritium are normally required to submit urine samples for bioassay on a periodic basis. The sampling period may be daily to biweekly or longer, depending on the potential for significant exposure. Usually, the period is weekly to biweekly.

Following an incident, or a work assignment with a higher potential for exposure, a special urine sample is usually required for each individual involved. The preferred method is to wait about 2 to 4 hours for the equilibrium to be established. The bladder is then voided. A sample submitted soon thereafter should be reasonably representative of the body water concentration. A sample collected before equilibrium is established will not be representative because of dilution in the bladder or because the initial concentration in the blood will be higher than an equilibrium value. However, any early sample may still be useful as an indication of the potential seriousness of the exposure.

At the bioassay laboratory, 1 ml of the urine is typically mixed with 10 to 15 ml of a suitable scintillation cocktail and counted in a liquid scintillation counter. At many laboratories, the urine is initially counted raw, and if the concentration is above a certain value (e.g., 0.1 µCi/l), the urine is distilled or spiked with a standard and recounted. The counting efficiency may be affected by quenching, although this can be corrected electronically.

The dose equivalent rate in the body water can be calculated directly from the concentration of HTO in body water, which until recently, was considered to be equivalent to the dose rate to the critical organ. ICRP-30 states that the average dose to the soft tissue could be taken to be equal to the effective dose equivalent. This change effectively dilutes the tritium, and thereby lowers the dose rate accordingly.

From this discussion, the dose equivalent rate, $R(t_0)$, to the soft tissue (63 kg), from a urine concentration of C_0 can be calculated as follows:

$$R(t_0) = C_0 \left(\frac{\mu Ci}{l}\right) \times 3.7 \times 10^4 \left(\frac{Bq}{\mu Ci}\right) \times 5.7 \times 10^3 \left(\frac{eV}{Bq}\right) \times 8.64 \times 10^4 \left(\frac{sec}{day}\right)$$

$$\times \frac{42\,l}{6.3 \times 10^4 \text{ grams}} \times 1.6 \times 10^{-12} \frac{erg}{eV} \times 10^{-2} \frac{rad\text{-}gram}{erg} \times 1.0 \frac{rem}{rad} \qquad (F.3)$$

$$= 1.94 \times 10^{-4}\, C_0\, \frac{rem}{day}.$$

From the dose rate R(t), the committed dose (D_∞) can be calculated from

$$D_\infty = \int_0^\infty R(t)\, dt. \qquad (F.4)$$

Following a bioassay measurement, the quantity R(t) can be estimated from an assumed biological half-life. A previously measured value (for that individual) or the average value (for reference man) of 10 days may be used. In that case,

$$D_\infty = R(t_0) \int_0^\infty e^{-\lambda t}\, dt = R(t_0) \int_0^\infty e^{-0.693t/T_{Bio}}\, dt$$

$$= \frac{R(t_0)}{0.693}, \qquad (F.5)$$

where, D_∞ = the committed dose equivalent,

$R(t_0)$ = the daily dose rate at $t = t_0$,

λ = the elimination constant, and

T_{Bio} = the biological half-life in days.

However, if a more precise calculation of the individual's dose is required, the actual biological half-life should be determined from the values of subsequent bioassay data.

For very low exposures (<1 to 10 μCi/l), no great error is incurred by assuming a constant half-life between weekly sampling points. For higher exposures, a greater sampling frequency is recommended to determine the dose more accurately.

As was noted above, a pure HT exposure can be thought of as a combination of a lung exposure from the HT and a whole body exposure from the HTO converted from the HT dissolved in the blood. The whole body dose can be determined as outlined above by analysis for HTO in the urine. Since the effective dose equivalents from the lung and whole body exposures are approximately equal, the total effective dose can be conservatively obtained by multiplying the HTO whole body dose by 2.

In general, this is too conservative (by the factor of 2), since a release of pure tritium gas with <0.01% HTO is highly unlikely. With only a slight fraction (~1%) of HTO in the air, the effective dose is essentially the HTO whole body dose as determined by bioassay.

In any exposure to HTO, a certain small fraction of the tritium will exchange with non-labile organic hydrogen in the body, there to remain until metabolism or exchange eliminates the tritium. Following a high acute or any chronic exposure, two- and three-component elimination curves have been observed (ranging from 30 to 230 days). Although most of the dose is due to the HTO in all of these observed cases, such exposures should be followed until urine concentrations are down to the range of <0.1 to 1 μCi/l, in order to calculate the dose more precisely.

It has also been observed that skin contact with metal surfaces contaminated with T_2 or HT produces tritium-labeled molecules in the skin (possibly catalyzed by the metal), which in turn results in longer elimination times for the labeled or metabolized constituents. Lung exposure to airborne metal tritides may also cause unusual patterns of tritium concentrations in body water, due, supposedly, to retention of these particulates in the lung with subsequent leaching and conversion to organically bound tritium. For these and other reasons, it is good practice to follow the elimination data carefully, and to look for organically bound tritium in the urine.

F.2.2 Dose Reduction

As was noted above, the committed dose following an HTO exposure is directly proportional to the biological half-life, which in turn is inversely proportional to the body-water turnover rate. This rate varies from individual to individual. As may be expected, such things as temperature, humidity, work, and drinking habits may cause rate variations. Although the average biological half-life is 10 days, it can be decreased by simply increasing fluid throughput, especially of liquids that are diuretic in nature, e.g., coffee, tea, and beer. The half-life may then be easily reduced to 4 to 5 days; however, a physician should be consulted before any individual is placed on a regimen that might affect his/her health. It is essential that medical supervision be involved if diuretics are taken because the resultant loss of potassium and other electrolytes can be very serious if it is not replaced. Such drastic measures may result in a decrease in half-life to 1 to 2 days. Even more drastic is the use of peritoneal dialysis or a kidney dialysis machine. These may reduce the half-life to 13 and 4 hours, respectively. Such techniques, although extreme, should be used only in life-threatening situations, involving potential committed dose equivalents that would exceed a few hundred rem without such treatment.

Individuals whose urine concentrations exceed established limits should be relieved from work involving possible further exposure to radiation, whether from tritium or other sources. Limits are generally suggested or imposed by the health physics organization to make certain that the annual worker dose limits are not exceeded. The operating group may impose even stricter limits on their staff than those imposed by the health physics group. The actual values, which may range from 5 to 100 μCi/l, are often dependent on the availability of replacement personnel, and the importance of the work that needs to be accomplished.

Results of bioassay sampling should be given to workers who submit samples as soon as they are available. The results may be posted or the workers may be personally notified. Moreover, the results are required to be kept in the workers' personal radiation exposure records or medical files. Like any other radiation exposure, any dose in excess of the regulatory limits must be reported to the appropriate authorities.

F.3   Measurement Techniques

Because an extensive review of tritium measurement techniques is beyond the scope of this document, it will be assumed that the reader is already acquainted with the fundamentals of radiation detection instruments. However, for those not familiar, an extensive review of *Tritium Measurement Techniques* can be found in the National Committee for Radiation Protection's (NCRP), NCRP-47.[F-4] Moreover, a review of site-specific measurement techniques can also be found in Department of Energy's (DOE's), WASH-1269, *Tritium Control Technology*."[F-5] The bulk of the following has been adapted from both sources. Since both documents were published in the 1970s, it can be expected that some of the information will be dated, although the basic measurement techniques have changed very little since that time.[‡]

This section discusses instruments or techniques used for monitoring tritium for health and safety purposes. However, since process-monitoring instruments often involve the same or similar detectors, they are also included in the discussion.

F.3.1   Air Monitoring

Ionization chamber instruments are the most widely used instruments for the measurement of tritium in gaseous (and vapor) forms in laboratory, environmental, and process monitoring applications. Such simple, economical devices require only an electrically polarized ionization chamber, suitable electronics and, in most cases, a method for moving the gas sample through the chamber, which is usually a pump. Chamber volumes typically range from a tenth- to a few tens of liters, depending on the required sensitivity. The output is generally given in units of concentration (multiples of μCi/m^3 or Bq/m^3), or, if a commercial electrometer or pico-ammeter is used, in current units which must then be converted to concentration. A rule of thumb that can be used to convert current to concentration is: concentration (μCi/m^3) = $10^{15} \times$ current (amps)/chamber volume (liters). For real-time tritium monitoring purposes, the practical lower limits of sensitivity range from 0.1 to 10 μCi/m^3.

For measurements of low concentrations, sensitive electrometers are needed. For higher concentrations, e.g., >1 mCi/m^3, the requirements on the electronics can be relaxed, and smaller ion chambers may be used. Smaller chambers also need less applied voltage, but because of a greater surface area to volume ratio, there is a greater likelihood for residual contamination in the chamber, which elevates the background. Response times for higher-level measurements can be made correspondingly shorter. However, small chambers and chambers operated at low pressures may have significant wall effects so that the above rule-of-thumb may not apply. Such instruments would have to be calibrated to determine their response.

[‡]   For more recent information on the measurement techniques used at various DOE sites, see also the references cited at the end of this section under the heading of *Suggested Additional Reading*, i.e., 1) *Primer on Tritium Safe Handling Practices*, DOE-HDBK-1079-94, December 2001; 2) *Radiological Training for Tritium Facilities*, DOE-HDBK-1105-96, December 2001; and 3) *Tritium Handling and Safe Storage*, DOE-HDBK-1129-99, March 1999.

Although most ionization chambers are of the flow-through type, that require a pump to provide the flow, there are presently a number of facilities that use so-called "open window" or "perforated wall" chambers. These chambers, which may employ a dust cover to protect the chamber from dust and other particulates, allow the air or gas to penetrate through the wall to the inside chamber. Such instruments are currently being used as single point monitors at several facilities for room, hood, glove box, and duct monitoring.

F.3.2   Differential Monitoring

Because of the greater toxicity of HTO compared to HT (25,000 times greater according to ICRP-30), it is often desirable to know the relative amounts of each species following a release into a room, or release to the environment. In the case of stack monitoring, this is more easily accomplished by taking discrete samples of the stack effluent using bubblers or desiccants in conjunction with a catalyst for oxidizing the HT (see Section F.3.3). For differential monitoring, the simplest technique is to use a desiccant cartridge in the sampling line of an air monitor. The result is a measurement of the HT concentration. Without the cartridge, the total tritium concentration is measured. Subtraction of HT from the total produces the HTO concentration. The technique may be used manually with one instrument or automatically by switching a desiccant cartridge in and out of the sampling line.

Another technique involves the use of a semi-permeable membrane tube bundle in the sampling line to remove the HTO (preferentially over the HT), which is then directed to an HTO monitor. After removing the remaining HTO with another membrane dryer, the sampled air is directed to the HT monitor. Although this technique is slower than the one requiring a desiccant cartridge (response and equilibrium times being 1 to 2 minutes and 10 to 20 minutes, respectively), it does not require a periodic cartridge replacement. Furthermore, it can be adapted to the measurement of tritium in both species in the presence of noble gases or other radioactive gases by adding a catalyst after the HTO dryers, followed by additional membrane dryers for the HTO converted from the HT by the catalyst.

F.3.3   Discrete Sampling

Discrete sampling differs from real-time monitoring in that the sampled gas (usually air) must be analyzed for tritium content by means of liquid scintillation counting (in the case of HTO). The usual technique is to flow the sampled air through either a solid desiccant (molecular sieve, silica gel, DRIERITE, etc.) or water or glycol bubblers. For low-flow rates (approximately 0.1 to 1 l/min), bubblers may be used. Bubblers are more convenient for sampling, but are less sensitive than the solid desiccant technique.

Glycol or water may be used, but glycol is generally preferred for long-term sampling. In any case, the collected water is then analyzed for HTO. For differential monitoring of HTO and HT, a heated catalyst (usually a palladium sponge) is used between the HTO desiccant cartridge or bubblers and the HT cartridge or bubblers. In a different arrangement, palladium is coated on the molecular sieve in the HT cartridge to oxidize and absorb the resulting HTO. This technique, however, is usually only employed for environmental monitoring.

Another technique for sampling HTO in air is to use a "cold finger" to freeze HTO out of the air; an alcohol and dry ice mixture in a stainless steel beaker works well. To arrive at the concentration, knowledge of the relative humidity is needed. A soft plastic bottle squeezed several times to introduce the air (containing the HTO) into the bottle is another method. A measured quantity of water is then introduced and the bottle is capped and shaken. In a minute or less, essentially all the HTO is taken up by the water, which is then analyzed.

Other techniques involve placing a number of vials or other small, specially designed containers of water, liquid scintillation counting cocktail, or other liquid in selected locations in the area being monitored. After a period of time (usually a number of days) the liquid in the containers is analyzed. The result is semi-quantitative (for open containers) to quantitative (for specially designed containers).

F.3.4 Process Monitoring

Ionization chambers are typically used for stack, room, hood, glove box, and process monitoring. The outputs can be used to sound alarms, activate ventilation valves, turn on detritiation systems, and for other functions. In general, it can be expected that stack, room, and hood monitors will require little non-electronic maintenance (i.e., chamber replacement due to contamination) because under routine circumstances, the chambers are constantly flushed with clean air and are not exposed to high tritium concentrations for extended periods of time. Glove box monitors, however, can be expected to eventually become contaminated, especially if exposed to high concentrations of HTO. Process control monitor backgrounds can also be expected to present problems if a wide range of concentrations (e.g., 4 to 5 orders of magnitude) are to be measured.

Mass spectrometers, gas chromatographs, and calorimeters are generally used as workhorse instruments for process monitoring. Because of their relative insensitivities, however, these instruments cannot be used for the detection of tritium much below a few parts per million (Ci/m^3). For this reason, care must be taken in the interpretation of analytical results and the related health physics concerns. It is not uncommon, for example, to find that samples that show no trace of tritium when analyzed on a mass spectrometer actually contain several curies of tritium.

F.3.5 Surface Monitoring

In general, it is not possible to measure the total tritium contamination on a surface except by destructive techniques. Even a slight penetration by tritium, for example, becomes quickly undetectable because of the weak energy of its beta particles. With open-window probes operated in the GM or proportional regions, it is possible to measure many of the particles emitted from the surface. However, quantifying that measurement in terms of the total tritium present is difficult since every exposure history is different, and the relative amounts of measurable to immeasurable tritium are consequently different. Such monitoring probes are then routinely used to measure the accessible part of the contaminating tritium. Care must be taken to protect the probe from contamination. When monitoring a slightly contaminated surface after monitoring a highly contaminated one, contamination of the probe can be an immediate problem. Placing a disposable mask over the front face of the probe can reduce but never eliminate this contamination completely, particularly when the tritium is rapidly outgassing from the surface being monitored.

For highly contaminated surfaces (>1 mCi/100 cm^2), it is possible to use a thin sodium iodide crystal or a thin-window GM tube to measure the characteristic and continuous x-rays (bremsstrahlung) emitted from the surface, as a result of the interaction of the beta particle with the surface material. In terms of total surface tritium, such measurements are semi-quantitative at best.

F.3.6 Liquid Monitoring

Liquid monitoring is almost universally done by liquid scintillation counting. For liquids other than water, care must be taken that the liquid is compatible with the counting cocktail. Certain chemicals can degrade the cocktail. Others are not miscible and may retain much of the tritium; still others result in a high degree of quenching. In addition, samples that contain peroxide, or that are alkaline, may result in chemiluminescence, which can interfere with the measurement. Such samples should first be neutralized before counting. Chemiluminescence and phosphorescence both decay with time, so that keeping the samples in darkness for a period of hours can usually eliminate the problem. Distillations may be necessary for some samples; use of quenching curves or a special cocktail may be necessary for others.

For rather "hot" samples, as may be the case for vacuum pump oils, bremsstrahlung counting may be useful. This technique may also be useful for active monitoring of "hot" liquids. Active monitoring of liquids may also be done with scintillation flow cells, which are often made of a plastic scintillator material, or of glass tubing filled with anthracene crystals. However, these flow cells are particularly prone to contamination by algae or other foreign material which can quickly degrade their counting efficiency.

F.4   Instrument Types and Calibration

Instruments used for monitoring tritium in air and on surfaces and for counting tritium samples are discussed in this section. Methods and sources for calibrating such instruments are also discussed. All instruments used for monitoring tritium for health and safety reasons should be calibrated regularly. The calibration frequency is typically 6-months for portable or other instruments receiving hard use, 12-months for fixed instruments, and 12 months or longer for simple instruments such as stack samplers.

F.4.1   Air Monitors

Ionization chambers that are used for air monitoring are described in Section F.3.1. The techniques used to calibrate ion chamber instruments can vary, but traditionally they are calibrated with tritium gas, if it is practical to do so. If an instrument (or an instrument system) is calibrated with tritium gas once, then it is generally not necessary to repeat that type of calibration. Thereafter, an electronic calibration from the front end of the electrometer preamplifier (if accessible) made with a calibrated current source (or calibrated resistor and calibrated voltage source) can be used. This is followed by a determination that there is adequate voltage on the chamber, and that the chamber is connected. The latter is verified by use of an external gamma source. Finally, if the chamber is of the flow-through type, proper flow must be verified.

Gas-flow proportional counters are not commonly used for air monitoring in the United States, although there has been some renewed interest in them in recent years. This type of instrument is common in West Germany where regulations require monitoring at very low levels. Advantages are enhanced sensitivity (approximately 0.01 pCi/m^3) and the ability to discriminate against background radiation. Disadvantages include: 1) increased cost and complexity, 2) need of a carrier-counting gas, 3) low-flow rate resulting in slower instrument response, and 4) limited range (up to approximately 1 mCi/m^3). Gas-flow proportional counters are particularly attractive as stack monitors, where increased sensitivity is desirable, and a slower response time is not a problem.

Liquid and plastic scintillation detectors have been developed in Canada and elsewhere to monitor for HTO in air, but apparently are not widely used for this purpose. The liquid scintillation counting technique is expensive because it requires a continuous supply of counting cocktail. The plastic scintillator technique, although not very sensitive, has some advantage with regard to size of the detector, which generally consists of two parallel plates of the plastic scintillator arranged in a flow cell. The scintillator, which is relatively insensitive to penetrating gamma rays, can be easily shielded from outside interference because of its small size. For instruments such as gas-flow proportional counters or scintillation counters, use of tritium gas for routine calibration purposes is probably more justified because of the nature of the detectors. This technique particularly applies to scintillation detectors because other techniques are not as effective in determining if the scintillation detectors are properly working.

F.4.2   Surface Monitors

Count rate instruments equipped with windowless gas-flow proportional probes, thin sodium iodide crystals, or thin-window GM tubes and used to monitor surfaces were described in Section F.3.5. Tritiated polystyrene sources can be used to calibrate survey instruments for surface monitoring. Sources are constructed of thin plastic disks for which the tritium beta emission rate from the surface can be determined and certified. The tritium counting efficiency of gas-flow proportional counters, under ideal conditions can approach 50%. However, normal conditions, i.e., dirty or porous surfaces can reduce the counting efficiency to 10% or less. More stable sources of ^{63}Ni can also be used to verify the operation of surface monitoring instruments. However, determination of the tritium counting efficiency cannot be simulated with ^{63}Ni.

F.4.3   Tritium Sample Counters

There are primarily two types of instruments for analyzing tritium samples for radiation protection purposes: gas-flow proportional counters, and liquid scintillation spectrometers.

Gas-flow proportional counters are commercially available with and without a window over the counting chamber and with and without a sample changer mechanism. Windowless counters should be used for tritium samples in order to obtain the maximum counting efficiency. When a large number of samples can be counted, a proportional counter with an automatic sample changer is recommended. When a number of samples need to be counted quickly, several proportional counters with single sample capacity may be used to obtain prompt results.

Tritiated polystyrene sources can be used to calibrate proportional counters for analysis of tritium samples. The tritium counting efficiency for 2π proportional counters can approach 50% under ideal conditions. However, when dirty smear papers or thick porous samples are counted, the counting efficiency may be reduced to 10% or less. More stable ^{63}Ni sources can also be used to verify the operation of proportional counters.

Detection with liquid scintillation counters has become established as the most convenient and practical way of measuring tritium in the liquid phase. Liquid scintillation counters are commercially available, many with capabilities for handling several hundred samples. The technique consists of dissolving or dispersing the tritiated compound in a liquid scintillator, subsequently detecting the light emitted from the scintillator, and counting the number of emissions. Major efforts in developing the technique have been directed to improving the detection efficiency of the photo-multipliers, distinguishing the tritium scintillation events from others, and in finding scintillator/solvent mixtures that can accommodate large volumes of sample (especially aqueous samples) without the degradation of the scintillation properties.

Liquid scintillation counters should be calibrated regularly by means of NIST-traceable standards. Quenching standards, often supplied by the manufacturer, may be used to establish the counting efficiency for tritium as a function of quenching ratio. The quenching ratio, and hence the counting efficiency, for individual samples can be determined routinely. The tritium counting efficiency for unquenched samples is usually about 35% to 50%.

F.5 Contamination Control and Protective Measures

Contamination control can be an effective method of limiting uptake of tritium by workers. In this section, smear surveys and off-gassing measurements are described as the primary methods of monitoring the effectiveness of contamination control. For situations where tritium contamination cannot be prevented, a number of protective measures are described that provide engineering controls over the spread of tritium contamination. Respiratory protection, gloves, and other protective clothing for working in tritium-contaminated environments are also described in this section.

F.5.1 Methods of Contamination Control

Any material exposed to tritium or a tritiated compound has the potential of being contaminated. Although it is difficult to quantify tritium contamination levels, there are several methods available to evaluate the existence and relative extent of contamination, including smear surveys and off-gassing measurements. Good housekeeping and work practices are essential in maintaining contamination at acceptable levels within a tritium facility.

The total amount of tritium surface contamination is not an indication of its health or safety implications. Rather, the loose, removable tritium is a more important indicator; this is the tritium that can be transferred to the body by skin contact, or that may outgas and become airborne. Loose contamination is routinely monitored by smears (or swipes), which are wiped over a surface and then analyzed for tritium content by liquid scintillation or proportional counting.

F.5.1.1 Smear Surveys

Surface monitoring by smear counting is an important part of the monitoring program at a tritium facility. It is used to control contamination, to minimize uptake by personnel, and to prevent, or minimize, its spread to less contaminated areas. A routine surface contamination-monitoring program is required, and additional special monitoring should be provided when the condition or situation is warranted.

An effective tritium health physics program must also specify the frequency of routine smear surveys. Based on operating experience and potential contamination, each facility should develop a routine surveillance program that includes daily smear surveys in areas such as lunchrooms, step-off pads, and change rooms. In other locations within a facility, it may be sufficient to perform weekly or monthly routine smear surveys. In addition to the routine survey program, special surveys should be made on material being moved from one level of control to a lesser-controlled area. This will help prevent the spread of contamination from controlled areas.

Smears are typically small round filter papers used dry or wet (with water, glycol, or glycerol). Wet smears are more efficient in removing tritium and the results are more reproducible, although the papers are usually more fragile when wet. However, tritium smear results are only semi-quantitative, and reproducibility within a factor of two agreement (for wet or dry smears) is considered satisfactory. Ordinarily, an area of 100 cm^2 of the surface is wiped with the smear paper and quickly placed in a liquid scintillation counting vial with about 10 ml of cocktail, or 1 or 2 ml of water with the cocktail added later. It is important to place the swipe paper in liquid quickly after swiping since losses by evaporation can be considerable, especially if the paper is dry. The counting efficiency is not much affected by the presence of a small swipe.

Foam smears are also commercially available. These dissolve in most cocktails and do not significantly interfere with the normal counting efficiency. Alternatively, the smear paper may be counted by gas-flow proportional counting but, because of the inherent counting delays, tritium losses prior to counting can be significant. Moreover, counting efficiencies may be difficult to determine and can be expected to vary greatly from one sample to the next. Another drawback is potential contamination of the counting chamber when counting very "hot" smears. For all of these reasons, a liquid scintillation counting is the preferred smear-counting system.

F.5.1.1.1 *Allowable Tritium Surface Contamination Levels—Background*

In the traditional sense, the Nuclear Regulatory Commission (NRC) has not had to deal with tritium contamination, and/or with allowable tritium surface contamination levels, as these historically have come under the purview of the Department of Energy (DOE), and/or its predecessor agencies, i.e., the Energy Research and Development Agency (ERDA) and, prior to ERDA, the Atomic Energy Commission (AEC). It is interesting to note, however, that the subject of allowable tritium surface contamination levels had fallen through the regulatory cracks for years, because, in spite of the existing ICRP dose models for allowable surface contamination limits for most other radionuclides, the ICRP models contained a disclaimer: "These data are not applicable to pure beta-emitters with a maximum energy equal to, or less than, 150 keV." As a consequence, allowable surface contamination limits for tritium, and carbon-14, simply did not exist.

Some of that began to change in 1977, when the ICRP published their latest recommendations for the safe handling of radioisotopes in hospitals and medical establishments.[F-6] In their publication of ICRP-25, the ICRP was suggesting a general purpose working limit of 1 nCi/cm^2 for allowable radionuclide contamination on surfaces. For tritium and carbon-14, however, ICRP-25 specifically noted that the 1 nCi/cm^2 recommendation could be increased by a factor of 100. Using the appropriate scaling factors, the ICRP-25 recommendations, therefore, were suggesting that the maximum limit for tritium and carbon-14 contamination control levels for controlled area usage should be on the order of 10 μCi/100 cm^2, or 2.22×10^7 DPM/100 cm^2.

In one of the earliest attempts to address the problem for unrestricted use, the State of California, as an Agreement State, adopted an interim set of tritium and carbon-14 surface contamination limits, in 1977,[F-7] based on the existing guidance provided in the AEC's Regulatory Guide 1.86.[F-8] For the most part, the limits went unquestioned, and, over the years, the same set of limits was adopted by the San Francisco Operations Office of the DOE.[F-9] Thus, for DOE, the allowable surface contamination limits for removable tritium were set at 10,000 DPM/100 cm^2.

Everything went reasonably well until 1989, when the DOE published its final version of DOE Order 5480.11.[F-10] Like the NRC had done with its *Table of Acceptable Contamination Levels* in Reg. Guide 1.86, the DOE had also published a comparable table of *Surface Radioactivity Guides*, in a simplified format, in DOE Order 5480.11.

But, what is important to note with respect to the DOE's first version of the Order, is that the DOE did not include a separate category for tritium (or carbon-14). As a consequence, the DOE tritium community found that its regulatory limits for allowable surface contamination limits had been unexpectedly, and arbitrarily, reduced by an order of magnitude. (Tritium was now considered as falling into a generic category, along with β-γ emitters and nuclides with decay modes other than α-emission or spontaneous fission.)

When the tritium community objected *en masse*, on both a national and international basis, the DOE established the Tritium Surface Contamination Limits Committee, to look into, and correct, the problem. Although the Tritium Surface Contamination Limits Committee came back with recommendations that were more on the order of 100,000 DPM/100 cm^2 for removable tritium surface contamination, the DOE elected to adopt a more conservative limit of 10,000 DPM/100 cm^2. (See References F-11 and F-12, respectively.) It is particularly important to note, however, that, while the DOE has used the value of 10,000 DPM/100 cm^2 for the free release of tritium contaminated items from controlled areas, the tritium surface contamination limits used by the DOE are intended primarily for use in occupational exposure situations, and *not* for the free release of tritium contaminated items to uncontrolled areas.

F.5.1.1.2 Allowable Tritium Surface Contamination Levels—Facility Issues

Because they have not had to deal with the issue in the past, there is no obvious reason to expect the NRC to have any current limits in place to establish action levels to be used by operating facilities, e.g., nuclear reactors, for tritium surface contamination limits for occupational exposures, nor should they be expected to have limits in place to address the subject of the free release of tritium contaminated items to uncontrolled areas. As a starting point, therefore, the adoption of the original recommendations of the Tritium Surface Contamination Limits Committee, i.e., 100,000 DPM/100 cm^2 for operational limits in controlled areas, and 10,000 DPM/100 cm^2 for the free release of tritium contaminated items to uncontrolled areas would be appropriate. From an operational standpoint, experience has shown that both values can be used without placing undo administrative burdens on the staff. More importantly, from a health and safety standpoint, the information contained in the Committee's Report[F-11] has shown that both values are extremely conservative, for both the workers, and the general public.

F.5.1.1.3 Allowable Tritium Surface Contamination Levels—Transportation Issues

Although the Department of Transportation (DOT) has no specific limits in place to address allowable *tritium* surface contamination, the requirements specified in 49 CFR 173.443(a) do address allowable surface contamination limits on the external surfaces of *all* radioactive material transportation packages. The basic limit specified, for all radionuclides, is that the allowable surface contamination limits, for non-fixed (removable) contamination, must be kept as low as reasonably achievable (ALARA). The limits further specify that the allowable surface contamination limits, for non-fixed (removable) contamination, for β-γ emitters, is 4 Bq/cm^2, 1×10^{-4} μCi/cm^2, or 220 DPM/cm^2, all of which translate, in more conventional units, to 22,000 DPM/100 cm^2. Given the background information noted above in Section F.5.1.1.1, such a value is well in keeping with tritium operational issues and expectations.

The allowable surface contamination limits on the internal surfaces of transportation packages are addressed in 49 CFR 173.428(c), where it is stated that, for an *empty* package, the internal surface contamination levels must not exceed 100 times the limits specified in 49 CFR 173.443(a), or 2.2×10^6 DPM/100 cm^2. For the shipment of packages that have been used previously for the shipment of irradiated TPBARs, such a value becomes problematic in that, as was noted in the main body of this document in Section 7.5.3, once a package has been used for the shipment of irradiated TPBARs, it can probably, never again, be shipped as an *empty* package.

F.5.1.2 Out-Gassing Measurements

Basic out-gassing measurements can be made using any of several different methods. The most reliable methods, however, involve the use of a closed-loop system of known volume, and a flow-through ionization chamber monitor. By placing the material inside the volume and by measuring the change in concentration over a period of time, accurate determinations of tritium off-gassing rates can be made on virtually any material. The initial out-gassing rate measured is the required value, since the equilibrium concentration may be quickly reached in a closed volume, especially if the volume is small. Relative health hazards can be determined in absolute terms and, where appropriate, decisions can be made regarding the release of such materials to uncontrolled areas.

F.5.2 Protection Against Airborne Contaminants

Several important engineering controls are available for tritium protection. For the protection of personnel against potential inhalation hazards from tritium, the most commonly used methods include differential pressure zoning, dilution ventilation, and local exhaust ventilation techniques. Depending on the relative hazard, however, additional measures must be considered. In order of increasing protection factors, these might include but are not limited to air-supplied respirators (self-contained breathing apparatus), air-supplied suits, and glove boxes.

F.5.2.1 Differential Room Pressure Zones

Differential room pressure zones are used in virtually all tritium facilities. In general, this technique establishes a natural flow path that leads from less to more hazardous areas. Used in conjunction with dilution ventilation and local exhaust ventilation techniques (see Sections F.5.2.2 and F.5.2.4, below), differential zoning is an important line of defense against the migration of tritium into areas where it is not wanted.

Typical pressure zoning controls should be arranged as follows:

- Using outside air pressure as the reference, office areas and other uncontrolled areas will generally be held between zero (0.00) differential and -0.01 inches of water column;

- Main access corridors outside of the radioactive materials area (RMA) will generally be held between -0.01 in. and -0.025 in.; main access corridors inside the RMA will generally be held between -0.01 in. and -0.05 in;

- Individual rooms within the RMA will generally be held between -0.1 and -0.15 in; and

- Working arrangements for glove boxes will typically range from -0.25 in. to -1.0 in., depending on the comfort level of the operators.

In special cases, the pressure differentials may differ from those in the above example.

F.5.2.2 Dilution Ventilation

Dilution ventilation is the once-through flow technique of exchanging outside air for inside air for purposes of comfort and basic contamination control. For comfort control, this technique typically uses cooled air in the summer and warm air in the winter. However, dilution ventilation techniques are inherently inefficient for saving on energy. For contamination control purposes, dilution ventilation techniques are made even more inefficient because large quantities of air are occasionally required for the adequate dilution of room air releases in relatively short time frames.

F.5.2.3 Room Air Exchange

Room air exchange rates in most working environments are typically set to about four air changes per hour. At most tritium facilities, however, exchange rates are routinely set to ten air changes per hour in radioactive materials areas and four to six air changes per hour in offices and other non-controlled areas. Thus, depending on the size of the facility, it can be expected that the total air throughput for any given tritium facility will be approximately 10^6 to 10^8 m^3/day, or higher. Because of increased energy costs in recent years, studies have been conducted at a number of sites in which the feasibility of retrofitting air-handling systems with computerized flow control systems has been examined. The newer systems would automatically cut back on airflow rates during non-peak periods, and/or when facilities are unoccupied. Although few systems have actually been installed and tested, the impact of such systems should be such that health physics programs will not be affected.

It is important for health physicists to know room air exchange rates to determine waiting times before re-entering a room after tritium releases. Assuming that air change rates are ten volume changes per hour, the formula that may be used to determine room tritium activity is

$$\text{Final Value} = \text{Initial Value} \times e^{-10t}, \tag{F.6}$$

where t is the total time in hours after the release. The initial value of tritium air activity is assumed to have reached equilibrium.

F.5.2.4 Local Exhaust Ventilation

The primary advantages of local exhaust ventilation techniques, effective in tritium facilities, relate to the complete capture of the contaminant, regardless of its evolution rate, relative toxicity, or physical state. In addition, these techniques use relatively low-air volumes compared to dilution ventilation. Potential disadvantages of local exhaust ventilation techniques are their relatively complex system design and that, once most systems are installed, they cannot easily be moved to other locations.

F.5.2.4.1   Fume Hoods

Fume hoods are often used in local exhaust ventilation systems. In theory, linear flow established at or near hood openings (face velocities) capture the contaminants and draw them through the hood and into the connecting ductwork. The capture of gases and vapors will generally require lower-face velocities than those needed for the capture of particulates. Large and intermediate-sized particles, for example, will sometimes be difficult to capture because of their inherent mass and the forces of gravity. Smaller particles, on the other hand, (below a few microns in size), can be expected to behave in a manner similar to that for gases and vapors.

For tritium work in a fume hood, face velocities in the range of 100 to 150 linear feet per minute (lfpm) are used. Higher velocities, e.g., 150 to 200 lfpm, can produce turbulent flow, resulting in eddy currents that can sweep tritium back to the operator. Since the problem can be further compounded by the location of equipment within the hood, operations involving the use of fume hoods should be periodically reviewed to ensure that adequate protection is being provided.

F.5.2.4.2   Canopy Hoods

Canopy hoods are used in place of fume hoods for housing large equipment. Designed for specific applications, canopy hoods are used at many tritium facilities for the following: 1) to enclose glove box pass-through-port operations, 2) to house many experiments which are too large to fit into a fume hood, and 3) in some applications, to house tritium gas pumping systems.

Canopy hoods, although used with either natural or forced air exhaust, are most effective for hot- and warm-air processes where rising thermal currents help pull air into the hood. For tritium work, canopy hoods are usually designed such that heat-producing equipment (e.g., pumps) can be placed at floor level. Hood door openings, which usually slide to the right and to the left, must be designed so that they can function without interfering with the worker or the operation. However, because the protection afforded by canopy hoods can quickly be lost when cross drafts are introduced, hood openings must be kept to a practical minimum whenever the hood is in use.

F.5.2.4.3   Recovery/Cleanup Systems

It is common in many facilities with glove box operations to clean up the air and remove or recover the tritium from the air prior to exhausting to the atmosphere. Various stripper systems and recovery units are used for this purpose. Since environmental concerns are increasing, it is important to maintain environmental releases ALARA.

F.5.2.5 Respirators

In general, respirators that are effective for tritium fall into two categories: air-purifying respirators and/or air-supplied respirators. Air-purifying respirators usually contain chemical cartridges, special filters, or both, which remove contaminants from air prior to breathing. Air-supplied respirators are of two types: 1) the self-contained type, for which a cylinder of air (or oxygen), or an oxygen-generating chemical provides the necessary oxygen for breathing, or 2) the hose-type respirators for which air is supplied from an external source. Although ANSI Z88.2[F-13] describes in detail the types of respiratory protection devices that are appropriate for various types of chemical and radiological hazards, the primary use respirators in a tritium facility is to provide protection against the possible inhalation of HTO. To be effective against HTO, however, respirators must be of the type to remove HTO from air, exchange it for normal water vapor, or be supplied with an external source of clean air.

F.5.2.6 Air-Supplied Suits

Because of the inherent disadvantages normally associated with respirators and other breathing apparatus, air-supplied plastic suits that completely enclose the body are widely used by facilities that process tritium. Prior to using air-supplied suits at DOE facilities, however, the suits must be tested and approved by a DOE Respirator Advisory Committee (RAC).[F-14]

The main objectives of air-supplied suits are to 1) provide a layer of circulating air between the worker and the suit, 2) provide an adequate supply of breathing air for the worker, and 3) maintain an adequate flow of air from the interior of the suit to the exterior to help keep the body cool. The incoming air must meet the criteria of Type 1, Grade D breathing air, as specified in the Compressed Gas Association Standard.[F-15] The air-supply system should be designed to ensure a high degree of reliability.

Capacity requirements for air-supply systems will be dependent on flow requirements for specific suit designs. There are a wide range of flow rates used in RAC-approved suits (from 6 to 20 cfm per suit), and it is not uncommon to have several workers on a manifold system at the same time. Therefore, system capacities should be designed to provide adequate flow to each suit user. Capacities in excess of several hundred cubic feet per minute may be needed per system.

For tritium work, air-supplied suits must be constructed of materials that have acceptable permeation protection against HTO. They must also provide appreciable tear and abrasion resistance. Because they are intended for use in many different environments, suits must be designed to provide adequate vision, to minimize interference with normal work movements, and to be put on and taken off easily. Noise levels in suits resulting from the flow of incoming air must be maintained at levels less than OSHA workplace standards, and they must comply with RAC criteria. Because of the closed environment, and because of the additional background noise caused by the flow of air into the suits, communication methods between personnel may require special equipment.

F.5.2.7 Temporary Enclosures

A more effective way to contain tritium may be to construct a tent, (temporary canopy hood or a temporary glove box). The primary difference between the two is that hoods generally exhaust to the stack and glove boxes generally exhaust to cleanup systems. For tritium, tents can be thought of as being the nominal equivalent of a reactor-type contamination control point when large pieces of equipment or entire areas must be worked on.

Structural members for tents can literally be anything. Smaller glove-bag operations, for example, recommend the use of Tinker-Toys® for support. For larger operations, PVC pipe, scaffolding supports, and standard off-the-shelf fittings can be used, along with anything else that is available. Tent walls are usually made of 3-, 6-, or 12-mil, fire-retardant PVC plastic sheeting, depending on strength requirements that may develop because of the facility's differential pressures.

Tenting operations are usually designed to allow personnel to work inside. In most cases, personnel working inside will wear air-supplied plastic suits. For these reasons, communication links between personnel inside and outside become vital. Moreover, because many tenting operations involve the use of welding, brazing, grinding, and/or other hot processes, additional emphasis must be placed on possible fire hazards.

F.5.3   Protection Against Non-Airborne Contaminants

The personnel protective equipment worn by workers is one of the most important aspects of an effective health physics program. Since tritium can be easily absorbed through the skin, or through inhalation, personnel protective equipment must protect against both exposure routes. The following describes protective measures and equipment that may be used for skin-absorption pathways.

F.5.3.1 Gloves, General

In some operations, the hands and forearms of workers can be exposed to high tritium concentrations in many forms, and the proper selection of gloves and glove materials is essential.

Many factors should be considered in selecting the proper type of glove. Factors to be considered in making the selection include chemical compatibility, permeation resistance, abrasion resistance, solvent resistance, glove thickness, glove toughness, glove color, shelf life, and unit cost. Gloves are commercially available in materials such as butyl rubber, natural rubber, neoprene rubber, neoprene and natural rubber blends, nitrile (Buna-N®), and polyvinyl chloride (PVC) plastics, polyvinyl alcohol (PVA) coated fabrics, and Viton®.

Table F-2 shows the chemical compatibility of eight of the available glove materials, along with recommended and non-recommended uses. The data clearly indicate that certain types of materials are not recommended for use with certain types of chemicals. Different types of gloves should be readily available for use in routine handling of chemicals.

Table F-2.   Chemical Compatibility of Available Liquid-Proof Gloves

Material	Recommended For	Not Recommended For
Butyl	Dilute acids and alkalies, ketonic solvents, gas and vapor permeation protection	Petroleum oils, distillates, and solvents
Natural Rubber	Ketonic solvents, alcohols, photographic solutions	Petroleum oils, distillates, and solvents
Neoprene	Concentrated nonoxidizing acids and concentrated alkalies	Halogenated or ketonic solvents
Neoprene/Natural Blends	Dilute acids and alkalies, detergents, and photographic solutions	Halogenated or rubber ketonic solvents
Nitrile	Petroleum-based solvents, distillates, and oils	Halogenated or ketonic solvents
PVC	General purpose, low-risk hand protection	Halogenated or ketonic solvents
PVA	Halogenated solvents, paint shop applications	Water or water-based solutions
Viton	Halogenated solvents, concentrated oxidizing acids	Aldehydes, ketonic solvents

Table F-3 lists some of the physical properties of commercially available gloves that can be found in common use at most facilities. Listed in order of their cost, prices can be expected to range from well under $1 per pair for the thinnest (0.005 in. thickness) PVC gloves to more than $30 per pair for Viton (0.012 in. thickness).[§] Also included in Table F-3 are additional considerations for glove length, as well as comparisons of shelf life, glove toughness, and HTO permeation characteristics.

The rating system for the data in Table F-3 is as follows: Under Shelf-Life, "Excellent" refers to an indefinite time span with no obvious loss of properties; "Poor" refers to a time span of between 6 and 12 months, the loss of basic properties being obvious; "Fair" and "Good" refer to arbitrary time spans of 2 and 4 years, respectively, with some loss of properties becoming evident over time. "Relative Toughness" is a combined heading based on inherent glove properties reinforced by thickness where appropriate. The data suggest, for example, that the overall rating for nitrile gloves should not change appreciably with increasing thickness since toughness is a property inherent in the glove. For PVC gloves, however, the ratings do change with thickness because the relative toughness of PVC gloves is primarily a function of the cross-sectional area of the glove-body wall. The ratings for protection against HTO permeation are listed relative to butyl and Viton gloves, both of which are rated as "Excellent." For all of these ratings, it is assumed that the gloves will be discarded before steady-state permeation of HTO (HTO breakthrough) can occur. In all cases, these ratings are dependent on the total thickness of the glove, i.e., the cross-sectional area of the glove-body wall.

Table F-3. Physical Properties of Commercially Available Gloves

Glove Material	Length (in.)	Thickness (in.)	Shelf Life	Relative Toughness	HTO Permeation Protection
PVC	11	0.005	Fair	Fair	Poor
PVC	11	0.010	Good	Good	Fair
PVC	11	0.020	Excellent	Excellent	Good
Neoprene/Natural Rubber Blend	14	0.020	Good	Good	Good
Neoprene	11	0.015	Excellent	Good	Good
Neoprene	18	0.022	Excellent	Good	Good
Natural Rubber	11	0.015	Poor	Fair	Good
Nitrile	13	0.015	Excellent	Excellent	Good
Nitrile	18	0.022	Excellent	Excellent	Good
Butyl	11	0.012	Excellent	Poor	Excellent
PVA(a)	12	0.022	Good	Excellent	Poor
Viton	11	0.012	Excellent	Excellent	Excellent

(a) As a coated, flock-lined fabric, the thickness of PVA gloves can vary by as much as ± 20%. Because the PVA coating is water soluble, other properties of PVA gloves can also be expected to vary, depending on their long-term exposure to moisture.

Additional gloves that might be considered are polyethylene gloves (11 × 0.00175 in.) and surgeon's gloves (11 × 0.006 in.). Other properties that might be considered include the availability of powdered versus non-

---

[§] Price estimates listed are in 1980 dollar estimates.

powdered gloves. The former are important when dexterity is needed; the latter are better suited for high-vacuum and ultra-high-vacuum work.

The use of two or more glove layers should be considered for complex chemical operations such as waste treatment and handling; also, for maintenance operations that might include the potential for exposure to a wide variety of chemical compounds, such as plumbing replacement operations on large-scale vacuum effluent capture systems that have been in tritium service for several years. Although basic protection schemes can be determined for most combinations of chemical species, the best gloves are combined of three layers of liquid-proof gloves and an underlying layer of absorbent glove material, i.e., a cotton glove liner. Different-colored layers for indicating which layers fail to meet protection requirements should also be considered. This further means of protection would prove beneficial for most workers, except for the small percentage of workers who are colorblind.

F.5.3.2 Lab Coats and Coveralls

Lab coats and coveralls (fabric barriers) are worn at various times in almost all tritium facilities. Lab coats are normally worn for the general protection of street clothes as part of the daily routine. For added protection, coveralls are sometimes worn instead of a lab coat when the work is unusually dusty, dirty, or greasy. However, in most cases, the protection afforded by lab coats and coveralls is little more than cosmetic.

Unless they are treated with water-resistant, or waterproofing, agents, open-weave fabrics, such as those normally associated with lab coats and coveralls, provide minimal barriers against the airborne diffusion of HTO. Moreover, it can be expected that the HTO protection that is afforded will be the result of straightforward mechanical factors: some of the HTO will become absorbed in the weave of the fabric, some will be trapped in air pockets between layers of fabrics, and some will be trapped in air pockets that separate the fabric layers from the skin. Perspiration levels near the skin surface, both sensible and insensible, can be expected to add an additional short-term dilution factor. For the most part, however, it can be expected that, unless lab coats and/or coveralls are changed often, approximately every 10 minutes or so, diffusion and dilution effects will quickly reach equilibrium in high HTO concentration operations, and all barrier effects will be nullified.

Waterproof and water-resistant lab coats and coveralls have been tested at various laboratories. In most cases, however, they are not recommended for everyday use because of the excessive heat loads inflicted on the worker. Many facilities prefer the use of open-weave fabrics for lab coats and coveralls and the use of an approved laundry for contaminated clothing. Other facilities have opted instead to use disposable paper lab coats and coveralls, exchanging the costs associated with a laundry for the costs associated with replacement and waste disposal.

F.5.3.3 Shoe Covers

Although shoe covers can provide protection factors that range over several orders of magnitude, the routine use of shoe covers in a tritium facility must be thoroughly weighed against actual need. Like lab coats and coveralls, shoe covers offer little protection against spreadable particulates and/or gases and vapors. As a general rule, shoe covers are not recommended for the control of spreadable contamination, except in highly contaminated areas, because good housekeeping, i.e., regular dusting, washing, and waxing of floors, provides better control over contamination spread. For localized contamination problems, such as those that might results from spills of tritium-contaminated liquids and solids, the use of liquid-proof shoe covers should be considered to prevent the spread of contamination.

F.6   Decontamination

Methods available for decontaminating materials are based on material composition and the extent of tritium contamination. Effective decontaminating agents include soap and water, detergents, bleach, alcohol, and Freon®. Since decontamination is often difficult, especially where surfaces are exposed to high concentrations of tritium for extended periods, tools and specialized equipment routinely used in process areas should be stored there for reuse.

Action levels should be established for the different tritium facility control zones to ensure that tritium contamination levels do not build up over time. For example, smearable limits for uncontrolled material release and clean areas at different facilities may range from 1,000-to-10,000 dpm/100 cm^2. Smearable limits in controlled zones may be much higher, but an effective health physics program should have procedural limits on the amount of smearable contamination permitted. When these action levels are exceeded, timely decontamination efforts should be initiated.

In spite of all the precautions normally taken, there may be occasional tritium contamination of workers. Effective personal decontamination methods include rinsing of the affected part of the body with cool water and soap. If the entire body is affected, a shower should be taken using soap and water as cool as can be tolerated. This will help keep the skin pores from opening, thus minimizing skin absorption.

F.7 Maintenance

Maintenance activities and operations sometimes require work to be done on equipment outside of a hood or glovebox environment. Several techniques are available for this type of operation such as close-capture methods and contaminant huts or tents. Taking advantage of localized crosscurrents, "snorkels" and "elephant trunks" used as flexible exhaust lines can be placed directly over or adjacent to the work to be performed. Face velocities of several thousand lfpm can be generated to aid in keeping off-gassing tritium away from the workers. (See Section F.5.2, above).

F.8 References

F-1.   U.S. Department of Energy, *Health Physics Manual of Good Practices for Tritium Facilities*, MLM-3719, December 1991.

F-2.   International Commission on Radiological Protection, Publication 30 (ICRP-30), Part 1, *Limits for Intakes of Radionuclides by Workers*, Part I and Supplements, Pergamon Press, New York, 1979.

F-3.   International Commission on Radiological Protection, Publication 23 (ICRP-23), *Report of Task Group on Reference Man*, Pergamon Press, New York, 1974.

F-4.   National Commission on Radiation Protection and Measurements, *Tritium Measurement Techniques*, NCRP No. 47, Washington, DC (1976).

F-5.   *Tritium Control Technology*, Edited by T. B. Rhinehammer and P. H. Lamberger, Monsanto Research Corporation Report, WASH-1269, Miamisburg, OH, 1973.

F-6.   Recommendations of the International Commission on Radiological Protection: *A Report of Committees 3 and 4 on The Handling, Storage, Use, and Disposal of Unsealed Radionuclides in Hospitals and Medical Research Establishments*, ICRP Publication 25, Pergamon Press, Oxford, 1977. See, in particular, Section 9, Paragraphs 175-177, pp. 27-28.

F-7.   Honey, D.D., "State of California Inspection Policy Memorandum #8, Revised Guidelines for Decontamination of Facilities and Equipment Prior to Release for Unrestricted Use," 30 June, 1977.

F-8.   U.S. Atomic Energy Agency, Regulatory Guide 1.86, *Termination of Operating Licenses for Nuclear Reactors*, June 1974.

F-9.   U.S. Department of Energy, San Francisco Operations Office Management Directive, "Requirements for Radiation Protection," SAN MD 5480.1A. CH. XI, Section 8, paragraph (g), Attachment 1, 15 June 1987.

F-10.   U.S. Department of Energy, *Radiation Protection for Occupational Workers*, DOE Order 5480.11, December 21, 1988.

F-11.   Surface Contamination Limits Committee Report, c. 1991.

F-12.   Title 10, Code of Federal Regulations, Part 835, Appendix D, from 58 FR 65485, Dec. 14, 1993, as amended at 63 FR 59688, Nov. 4, 1998.

F-13.   American National Standards Institute, *American National Standard for Respiratory Protection*, ANSI Z88.2, 1988.

F-14.   O.D. Bradley, Acceptance-Testing Procedures for Air-Line Supplied Suits, LA-10156, Los Alamos National Laboratory, 1984.

F-15.   Compressed Gas Association, *Compressed Air for Human Respiration*, CGA-G7.1, New York, NY, 1968.

F.9   Suggested Additional Reading

U.S. Environmental Protection Agency, Federal Guidance Report #13, *Cancer Risk Coefficients for Environmental Exposure to Radionuclides: CD Supplement*, EPA 402-C-99-001, Rev. 1, Oak Ridge National Laboratory, Oak Ridge, TN, 2002.

Peterson, S-R. and P.A. Davis, "Tritium Doses from Chronic Atmospheric Releases: A New Approach Proposed for Regulatory Compliance," *Health Physics*, **82**, pp. 213–225, 2002..

U.S. Department of Energy, *Primer on Tritium Safe Handling Practices*, DOE-HDBK-1079-94, December 2001.

U.S. Department of Energy, *Radiological Training for Tritium Facilities*, DOE-HDBK-1105-96, December 2001.

U.S. Department of Energy, *Tritium Handling and Safe Storage*, DOE-HDBK-1129-99, March 1999.

NRC FORM 335 (9-2004) NRCMD 3.7	U.S. NUCLEAR REGULATORY COMMISSION	1. REPORT NUMBER (Assigned by NRC, Add Vol., Supp., Rev., and Addendum Numbers, if any.)
BIBLIOGRAPHIC DATA SHEET *(See instructions on the reverse)*		NUREG-1609 Supplement 2

2. TITLE AND SUBTITLE

Standard Review Plan for Transportation Packages for Irradiated Tritium-Producing Burnable Absorber Rods (TPBARs)

3. DATE REPORT PUBLISHED	
MONTH	YEAR
February	2006

4. FIN OR GRANT NUMBER

J5509

5. AUTHOR(S)

Ronald S. Hafner, Jason L. Boles, Chad L. Goerzen, Jack Hovingh, Gerald C. Mok, Edward W. Russell Jr., and Chol K. Syn

6. TYPE OF REPORT

Technical

7. PERIOD COVERED *(Inclusive Dates)*

8. PERFORMING ORGANIZATION - NAME AND ADDRESS *(If NRC, provide Division, Office or Region, U.S. Nuclear Regulatory Commission, and mailing address; if contractor, provide name and mailing address.)*

Lawrence Livermore National Laboratory
7000 East Avenue
Livermore, Ca. 94550-9234

9. SPONSORING ORGANIZATION - NAME AND ADDRESS *(If NRC, type "Same as above"; if contractor, provide NRC Division, Office or Region, U.S. Nuclear Regulatory Commission, and mailing address.)*

Spent Fuel Project Office
Office of Nuclear Materials Safety and Safeguards
U.S. Nuclear Regulatory Commission
Washington, DC 20555-0001

10. SUPPLEMENTARY NOTES

R.W. Parkhill, NRC Project Manager

11. ABSTRACT *(200 words or less)*

The NRC contracted with LLNL to compile this supplement to NUREG-1609 to incorporate additional information specific to tritium-producing burnable absorber rods (TPBARS). As a supplement to NUREG-1609, this report is intended to provide details on transportation package review guidance for the shipment of TPBARs. The principle purpose of this supplement is to ensure the quality and uniformity of staff reviews of packagings intended for transport of TPBARs. It is also the intent of this plan to make information about regulatory matters widely available, and improve communications between the NRC, interested members of the public, thereby increasing the understanding of the NRC staff review process. In particular, this supplemental guidance, together with NUREG-1609, assists potential applicants by indicating one or more acceptable means of demonstrating compliance with the regulations.

12. KEY WORDS/DESCRIPTORS *(List words or phrases that will assist researchers in locating the report.)*

TPBAR
Tritium
Tritium-Producing Burnable Absorber Rods
Tritium Rods
Spent Fuel
Standard Review Plan
SRP
Transportation Packages
Radioactive Material

13. AVAILABILITY STATEMENT

unlimited

14. SECURITY CLASSIFICATION

(This Page)

unclassified

(This Report)

unclassified

15. NUMBER OF PAGES

16. PRICE

NRC FORM 335 (9-2004)

PRINTED ON RECYCLED PAPER

www.ingramcontent.com/pod-product-compliance
Lightning Source LLC
Chambersburg PA
CBHW081552170526
45166CB00009B/2678